ATLAS
of poetic
ZOOLOGY

© 2019 Massachusetts Institute of Technology

Originally published as *Atlas de zoologie poétique.* © Flammarion, Paris, 2018.

Design by Karin Doering-Froger

This book was set in DTL Paradox by The MIT Press. Printed and bound in Portugal.

Library of Congress Cataloging-in-Publication Data

Names: Pouydebat, Emmanuelle, author.
Title: Atlas of poetic zoology / Emmanuelle Pouydebat ; illustrated by
    Julie Terrazzoni ; translated by Erik Butler.
Other titles: Atlas de zoologie poétique. English
Description: Cambridge, MA : The MIT Press, [2019]
Identifiers: LCCN 2018046310 | ISBN 9780262039970 (hardcover : alk.
    paper)
Subjects: LCSH: Zoology—Atlases. | Animals—Atlases.
Classification: LCC QL46 .P6813 2019 | DDC 590.22/2—dc23 LC record
    available at https://lccn.loc.gov/2018046310

10 9 8 7 6 5 4 3 2 1

ATLAS
of poetic
ZOOLOGY

EMMANUELLE POUYDEBAT

illustrated by
JULIE TERRAZZONI

translated by
ERIK BUTLER

THE MIT PRESS
CAMBRIDGE, MASSACHUSETTS
LONDON, ENGLAND

CONTENTS

To Alexandre, my extraordinary little animal,
an observer and experimenter who now, finally,
can leaf through a book by his mom!

To my brother Anthony and my parents
Gigi and Franck.

HOW WAS I to go about choosing thirty-six extraordinary animals when there are over a million species in existence? Where would I even start, when these miracles of nature, each in its own way, inspire such affection, wonder, and admiration in me? What are the criteria in the first place? Beauty? Strength? Behavior? Shape? Ability to adapt? Scarcity? Utility for human beings? Resemblance to us? Difference from us? Ancient origins? Strangeness? Risk of extinction? How could I select the marvels to include in this *Atlas of Poetic Zoology*?

After all, in the animal world everything is poetry. This word comes from the ancient Greek ποίησις (*poiesis*); the verb ποιεῖν (*poiein*) means "to make, to create." Animals are lyric poets; they discover and shape the world when they sing, dance, explore, and reproduce. Their poetry prompts us to reflect on how we see the world and all its inhabitants. The book at hand is an expression of wonder and a call to experience it for oneself.

Today, more than 1,211,612 animal species have been discovered and many researchers estimate that there are more than 8 million animal species in all. Animals have made everywhere their home: land, water, mountains, and the air; they live both at high altitudes and at abyssal depths, in arid deserts as well as wet regions, and places where it's extremely cold. From sea sponges to elephants, from ladybugs to human beings, each species is unique, even though they all share common ancestors. On this score, the oldest animal fossils seem to be from 700,000 million years ago. Holes most likely dug by bilaterians have been preserved from the Precambrian Era, and the remains of fauna from the Ediacaran Period include creatures like jellyfish.

No poetry is more beautiful than knowing that the animal world has been there for at least 700,000 million years, evolving all the while, and that we have the good fortune of being able to observe it and trying to understand its mysteries. The book at hand seeks to lead the reader to a land of wonder where turtles fly under the sea, lizards walk on water, insects impersonate flowers, birds don't fly, frogs come back from the dead, and virgin sharks give birth. Welcome to my world, where extraordinary beings feed, seduce, heal, revive, resist, protect, and defend—in brief, live and thrive.

UNBELIEVABLE ANIMALS

AN EXCEPTIONAL STANDARD

....

THE AFRICAN BUSH ELEPHANT
Loxodonta africana
Size: 3–4 m tall

....

NO ANIMAL inspires as much affection, fear, and admiration in me as the biggest land mammal alive. It's hard to find a species as complex and fascinating, or as paradoxical, as the African elephant. Complex and fascinating? Indeed! The elephant displays a broad array of incredible behaviors and physical aptitudes, some of which we cannot explain at all. Paradoxical? But of course. The elephant is so light when swimming, so gentle and patient with its young, and so selfless among its peers—yet so mighty and fearsome when charging (all five to seven tons of it) that no predator, apart from human hunters bearing arms, will risk challenging this beast. It's also paradoxical in terms of its evolution. Just think: the elephant's ancestor (*Eritherium azzouzorum*) was a tiny animal that lived about 60 million years ago, barely weighed five kilograms, and stood less than 50 centimeters at the shoulder! In the course of their history, elephants have also known the late, great mammoth (*Mammuthus sp.*), which saw the day some four million years ago; some were over twelve tons and five meters at the shoulder—twice as large as the biggest elephants now!

Clearly, the first outstanding trait of African elephants is their legendary memory, which helps them navigate great distances. African Elephants consume vast quantities of food; as a result, they're always moving from one place to the next. In five months, elephants in the Namibian desert can cover more than 600 kilometers. During the dry season, they scout out sources of water—as far as sixty kilometers apart—every four days. When there's a drought, elephants must leave their home range and try to find resources for the group to survive, especially its smaller members. Under such extreme conditions, the matriarchs—who possess excellent capacities for finding their way and remembering various stopping points—guide the others. In other words, only herds

"The elephant dies, but its defenses remain."

—*African proverb*

....

12

governed by mature females (35 years) with a lifetime of experience are able to leave familiar terrain and survive. At Amboseli National Park, Kenya, the elephants even commit the olfactory, visual, and vocal characteristics of nearby human beings to memory, so they can tell whether they're potential enemies (Maasai hunters, for instance) or simply tourists. Finally, elephants' sense of hearing is unbelievably sophisticated; they can pick up on low-frequency sounds humans don't even notice. This enables them to detect seismic signals in the ground and alert each other, across great distances, to the presence of a watering spot or predators. They can even sense rainfall more than 100 kilometers away. Thus, by listening to infrasound signals from clouds, they're able to steer their course toward downpours days, or even weeks, ahead.

African elephants don't just rely on their memory, either: they use tools. To get rid of flies, they rub themselves with branches and teach their young to do the same. They also dig holes to find water, then cover them up with bark to prevent it from evaporating. Elephants practice medicine; they pick out and eat plants and roots to heal themselves and induce childbirth. And here's the ultimate proof of their intelligence: they know how to play tricks! Asian elephants that have been domesticated stick mud in the bells that people put around their necks, to prevent them from ringing. It's the perfect way to avoid being found out when they decide to raid the fields!

These giants also feel empathy. It's impossible not to marvel at these colossal beings offering each other comfort, adopting orphans, working together to pull calves out of the mud, or pulling a spear from the flank of a wounded comrade. Even though we can't explain the practice fully, elephants observe funeral rites, too. They're aware of death and respect the remains of the departed. When one of them dies, they're upset and grieve; for several days, they group together and watch over the dead, mournfully sniffing and touching the body. In some cases, we can see this imprinting phenomenon and attachment to a deceased parent. But although elephants do gather around when one of their companions is about to die, the "elephant graveyards" of legend don't really exist. Large quantities of remains found in the same place might indicate a former watering hole, or represent the result of poaching.

In my eyes, elephants inspire profound humility and respect like no other animal. And deep sadness, too, when I consider the intolerable fact that the largest land mammal is in danger. It's estimated that one elephant is killed every fifteen minutes for the ivory trade.

Over time, baby elephant has grown up. The story has unfolded over more than 60 million years. The elephant has bested droughts and ice ages, and can even hear the noise made by clouds. And yet, in just half a century, we may have sealed the elephant's doom. This poem is a paradoxical mix of beauty, strength, and tragedy.

A SCALY, TOOTHLESS MAMMAL

THE PANGOLIN
Manis sp.
Size: 30–100 cm

WHO WOULD HAVE THOUGHT a stocky animal, about a meter long, toothless, and covered with scales on its back and tail—an animal that's usually nocturnal, lives on the ground or in the trees, and has a long tongue for drinking and eating termites and ants—could be a ... mammal? Biologists! Why? Because this creature, like all other mammals, is a viviparous vertebrate with teats and hair on its muzzle, belly, inside its legs, and between its scales. It's a warm-blooded animal; its so-called homeothermic physiology allows it to maintain a constant internal temperature between 37 and 40°C (98.6 and 104°F).

Having had the good fortune to observe it closely, I can declare without hesitation that the pangolin is a breathtaking animal. This is one creative beast. When it needs to defend itself, it curls up into a ball and tucks its head between its legs to protect itself from attacks by large carnivores like leopards and tigers. Getting it to unroll requires considerable effort! Humans can't pull it off. The gripping power in the pangolin's paws and tail (which represents up to 65 percent of the animal's total body length) is impressive. I experienced it firsthand when a female pangolin didn't want to leave my arm, which she'd decided to roll up around. Try telling a pangolin who's feeling comfortable that you'd like to go home. It's not easy ... the only solution is patience.

"Pangolin, you're my mother's brother, and you dig burrows that reach far away."
—*African proverb*

Besides its formidable power to hold tight, the pangolin can use its tail to help itself walk on two legs; plus, when it feels like it, this animal swims well.

The pangolin has another standout feature: it's an insect-eating machine, equipped with a long, protractile tongue up to 40 centimeters long (in the giant pangolin) that's covered with viscous saliva to which its prey sticks. An adult can consume several thousand ants in a single day, and some 70 million insects in a year! The pangolin has thick eyelids and hermetic valves in its nose and

ears; all this, together with the scales, offers protection from bites and stings. Pangolins don't chew with their mouths; instead, their stomach lining is covered with "horned teeth" for grinding up and digesting insects. And as if that weren't enough, pangolins swallow small pebbles to make sure their meals break down properly. What an amazing creature!

Unfortunately, poaching and illegal trade have made the pangolin the most endangered mammal in the world. It's hunted more than elephants and rhinos combined: each year, between 500,000 and 2.7 million pangolins are caught in the forests of Cameroon, the Central African Republic, Equatorial Guinea, and Gabon. In Africa, its scales are reputed to have magical powers. In Asia, its meat is a delicacy, and its scales—even though they're made of keratin, like human hair and nails—are thought to enhance virility, improve the health of nursing women, and possess healing properties. As a result, the pangolin is now in critical danger of extinction.

Among the Lega, a people in the Democratic Republic of Congo, the giant pangolin is a cultural hero, and it's forbidden to hunt it. The giant pangolin is thought to have first taught human beings how to build houses. As such, this majestic mammal stands at the center of a complex ritual affirming the maternal line in the human social order.

THE INSECT
THAT THINKS IT'S A FLOWER

THE ORCHID MANTIS
Hymenopus coronatus
Size: 6 cm in length (female), 3 cm (male)

THE POETRY OF ANIMALS often displays its greatest cruelty among the insects. On this score, the orchid mantis has achieved the heights of true artistry. What's more beautiful and more poetic than an insect fusing with a flower? And not just any flower—this mantis's legs look so much like the petals of an orchid, it's impossible to tell where one stops and the other starts. Its silhouette is incomparably subtle. The illusion is perfect. The orchid mantis provides one of the most impressive and aesthetically complete examples of mimicry. In viewing this animal as a flower, we all but forget that it's also a formidable predator—one with a marked preference for creatures every bit as poetic and beautiful: butterflies. It's beautiful but dangerous. Some mantises even catch small birds and bats. And though they look like parts of a flower, the front legs—"raptorial" limbs—are covered with thorns for seizing prey more effectively.

"The praying mantis catches the grasshopper, but the sparrow is watching from behind."
—Chinese proverb

In terms of strategies for self-protection or hunting, it doesn't stop there. The orchid mantis's young can already jump quickly and play dead; this is what's known as *thanatosis*. Once it's grown, a mantis can turn its triangular head more than 180 degrees and achieve almost periscopic mobility thanks to its rigid neck. In this way, it remains motionless while enjoying an extraordinarily broad field of vision and waiting for prey to arrive.

In closing, a quick word to reassure gentlemen-readers: the male orchid mantis mates by assuming a position at the very end of the female's abdomen. It's hardly a minor detail: doing so prevents him from being devoured by his partner …

THE LIZARD
THAT RUNS ON WATER

....

THE JESUS CHRIST LIZARD
Basiliscus plumifrons
Size: 70–90 cm

....

RUNNING ON WATER IS IMPOSSIBLE—but only if you don't count the plumed basilisk, nicknamed the Jesus Christ lizard. Supernatural powers? No … The plumed basilisk boasts a handsome locomotor adaptation: racing on its hind legs, it can move over the surface of the water without sinking—it's a little speeding bullet.

Up in the trees of dense tropical forests, plumed basilisks perch on branches overlooking rivers and streams. Although snakes and human beings pose a threat, the greatest danger comes from above. Predatory birds expect them to hide in the foliage. In fact, this lizard does just the opposite when it's being hunted. It drops onto the water and starts running at a pace of about ten kilometers an hour, giving it precious time to escape—and on two legs, at that! Remarkably, newborn plumed basilisks are already able to pull off this feat. It's not so surprising given that their own parents might eat them; they need to be quick from the start. It doesn't matter whether a plumed basilisk weighs two or two hundred grams: they can all run on water.

How is this exploit even possible? First, the basilisk puts its feet on the water so rapidly and so forcefully that they never sink more than a couple centimeters. A human being would have to race at more than 100 kilometers an hour to do the same. That said, it's not just a matter of velocity; in fact, these lizards run at a slower rate than their counterparts on land with the same size and shape (14 kilometers per hour), such as the zebra-tailed lizard. But if speed isn't the only issue, what else enters the equation? The dynamic equilibrium represents a biomechanical compromise between the lizard's low mass, the area of contact on the water, and an ingenious system of lateral movements performed by the hind legs in conjunction with the tail. The latter—which forms two-thirds of the lizard's body—plays a key role by

"A late riser doesn't see the lizard brushing its teeth."

—Maasai proverb

....

22

providing a counterweight; in addition, the tail strikes the water to create a kind of carrier wave. So is the Jesus Christ lizard a runner or a surfer?

At any rate, it's a fact that nature forces terrestrial vertebrates to deal with various material substrates: ground sloping at various angles (with any number of textures), branches of assorted dimensions pointing this way and that, surfaces that can be smooth or rough, and so on. Among all extant species—which move in so many environments and in so many different ways—the plumed basilisk is the only one that can cross a body of water, whether it's just hatched or has reached adulthood.

If a human being were to perform the same feat, he or she would have to run at a speed of more than 100 kilometers per hour—and have muscles fifteen times stronger than normal! The Jesus Christ lizard offers a wonderful example of all we can learn from nature: healthy modesty, combined with unbelievable abilities that, for the most part, are still probably unknown to us. It's time to open our eyes.

THE FISH
THAT CAN WALK

....

....

AFTER A LIZARD that runs on the water, here's a fish that walks under it! The party in question is the seadevil. This curious creature lives in the waters of the Galapagos. From above, its pectoral fins look like bat wings (hence the name), and it has a circular or box-shaped head. *Ogcocephalus darwini* is a remarkably strange animal, even within its own family, *Ogcocephalidae*. Between its flat, triangular body covered with bumps and spines, the "lipstick" it wears, its four "legs" and big "snout," and the small, retractable trunk it wields, it's hard not to think that somebody's pulling our leg.

Known for its almost fluorescent red lips—which likely enhance species recognition during the breeding season—the batfish isn't a terribly accomplished swimmer. It does a better job "walking" on the seabed, wandering around on the ends of its pectoral and pelvic fins. Because this musculature is strongly developed, it looks like it has stocky legs! These limbs even have a kind of fleshy pad at the outermost point. To walk, the creature doesn't need solid ground—or even feet!

Besides an unexpected capacity for locomotion, the batfish has the added feature, when hunting, of being able to use its *illicium* (from the Latin *illicere*, "to entice"). What's that, you ask? Mature batfish have a dorsal fin jutting out from the top of the head; it looks like a swollen nose or proboscis. The appendage serves as a means of attracting prey—small fish, crustaceans, and mollusks. But that's not all. Between the base of this protuberance and the fish's mouth is a surprising organ (most commonly hidden inside the body): a retractable lure! It's like a fishing rod that occasionally secretes a liquid to act as chemical bait. The batfish waves

"The sea never buys fish."

—*Turkish proverb*

....

it around to attract prey. Indeed, this same lure seems to provide a means of detecting potential meals. Needless to say, it's quite the setup—especially given that this animal

has other ways of hunting. Some cover themselves with sand or dig a hole to hide in. Such adaptations allow them to work stealthily, from a spot under the seafloor or on a rock, beckoning to guileless prey. As if those red lips weren't enough to attract unsuspecting victims!

Batfish are relatively understudied in
comparison to other fish. It's probably their
good luck: these little animals, far away
from human beings, are not endangered at the
moment.

A UNICORN OF THE SEA

....

THE NARWHAL
Monodon monoceros
Size: 4–5 m

....

DWELLER OF THE ICY ARCTIC depths, the narwhal is one of the most unfamiliar aquatic animals to human beings. This exceptional marine mammal is a cetacean that has inspired many myths. It's the only animal in the world with an immense, spiraling tooth that extends like a horn, whose function continues to pose questions to scientists. Even today, the narwhal is likened to the unicorn, the legendary medieval creature symbolizing spiritual force, purity, and magical power.

In the sixteenth century, Queen Elizabeth I of England is said to have paid 10,000 pounds for a narwhal's horn—the price of a castle! At the time, these armaments were reputed to neutralize the effects of poison. Because the narwhal is so rare and lives so far away, the legend persisted until the eighteenth century.

For five hundred years, the myth of the narwhal has thrived thanks to this mysterious anatomical feature. By turns, it's been considered a harpoon for hunting, an acoustic amplifier, a means of thermal regulation, a rudder, a weapon against rivals and predators, a tool for poking through the ice to breathe, and a secondary sexual characteristic for attracting females and establishing social hierarchy; the Inuit believe it holds magical properties. At any rate, we're now certain that this remarkable feature is a canine tooth twisting through the upper lip; in males, it can measure up to three meters (in contrast, the maximum length in females is 33 centimeters); the animal itself can grow to be four or five meters long. A recent study has shown that about ten million nerve endings in the core of the tooth connect the animal's brain to its ocean surroundings. In this capacity, the horn serves as a hydrodynamic sensor to detect the water's salinity, temperature, and pressure, giving the narwhal information about the position of ice, location of fish, and when it's time for migration. Images taken

> *"I'm a unicorn.*
> *I don't believe*
> *in human beings."*
> —*Anonymous*

by two drones in Tremblay Sound, in Canada's northernmost region, have also confirmed an old hypothesis: they show a narwhal harpooning a cod with its horn, then eating its prey! Clearly, then, this feature performs a number of functions: breaking through blocks of ice, serving as a weapon when confronting other males over a female, and even acting as a kind of radar. But mysteries remain. About 15 percent of females have a horn, but they have been little studied in comparison to males. The unicorn of the sea still harbors many secrets.

The narwhal's horn could almost make one forget its other peculiarities. This animal is an excellent diver; when hunting squid, octopus, crustaceans, mollusks, and fish, it can descend to a depth of 1,500 meters and stay there for half an hour looking for food. A single narwhal can perform up to twenty-six dives a day. Narwhals also emit a range of different sounds in order to communicate and navigate. Echolocation provides orientation and indicates where prey is to be found—the same as for bats. In addition, it seems that each individual has a unique voice, which helps members of a group know where their friends are. Narwhals are migratory animals; their summer territory covers about 10,000 square kilometers, and their winter territory measures some 26,000 square kilometers. These animals travel in keeping with ice-masses building up and melting. Needless to say, global warming and the vanishing polar ice caps pose a threat to this living myth.

*Hunted by polar bears, orcas, and human beings, the
narwhal is especially threatened by global warming,
which changes the temperature and salinity of the water.
Likewise, when the water surface freezes, narwhals
can become trapped. Human activities such as fishing,
boating accidents, and pollution from mining affect their
birth rates. Given that the female delivers a calf only
once every three years, we have reason to be worried.*

ZOANTHUS BOUQUETS

....

"RADIOACTIVE DRAGON EYE" CORAL
Zoanthus sansibaricus
Size: 1–1.5 cm (buccal disc)

....

HERE'S AN EXTRAORDINARY ANIMAL like a pretty bouquet of aquatic flowers bursting with color. Known as "eagle eye coral"—or, alternatively, "radioactive dragon eye coral"—this mysterious zoanthid belongs to the Cnidaria phylum. They're delicate animals who live in colonies, with bodies made up of small discs lined with tentacles; when light falls on them, they display a rainbow of phosphorescence ranging from green to blue, with orangish tints in between. Other shades can appear in green or blue concentric circles: muted blue, grayish hues, dark green, and so on. They live on sandy, muddy, or rocky surfaces on the seafloor. These magnificent animals are just as beautiful as they are fundamental to life on our planet. Over the course of 250 million years, corals have come to constitute one of the largest and most complex ecosystems on the globe. Corals are the lungs, the tropical forest, of the ocean: they absorb about half of the world's CO_2 emissions. What's more, over a million species of plants and animals live in close conjunction with them. More than 4,000 kinds of fish and thousands of plants and other animals dwell in the reefs they form. Corals protect coasts and coastal life from storms, waves, floods, and erosion … they're absolutely vital.

When something foreign touches them, their lining rolls up to afford cover. Like other animals, these soft corals have a digestive cavity, a nervous system, and muscle and urticating cells (*myocytes* and *cnidocytes*) to protect themselves. They produce a host of little molecules in order to avoid being eaten and ward off parasites and infection. And if things get serious, they can do even more. When exposed to open air, the radioactive dragon eye takes revenge by emitting a dangerous aerosol—a real poison. Dermal contact transmits palytoxin, a powerful vasoconstrictor that acts quickly. Symptoms include sensory disturbances (especially taste and touch), hypertension, respiratory distress, muscle pain, and

> *"Coral deep under the ocean's waters: an experience deep at the heart of adversity."*
> —Félix Bogaerts

even coma leading to death. Corals' chemistry has inspired respect; traditionally, fishermen steer clear of them.

To be sure, they produce toxins. But nature always holds surprises in store, and these corals are no exception. Specifically, they bring forth alkaloids—that is, the molecules that constitute the active ingredients of drugs. As such, zoanthids represent a potential treasure trove for helping humankind. Furthermore, they synthesize chloroform methanol. What's the big deal? Well, it so happens that this chemical compound is capable of destroying a parasite that invades the human lymphatic system! The substance even sterilizes female attackers. Other chemicals made by members of the Zoanthus genus—such as norzoanthamine—can prevent the loss of bone density and may therefore play a key role in developing drugs to treat osteoporosis. Flower-animals that heal … is there anything more beautiful to protect?

Gardens of the sea, lungs of the ocean:
they're crucial for life. It's easy to appreciate,
then, how dire the situation is. Basically,
all coral reefs on our planet are threatened.
So beautiful, powerful, and fragile—all at
once—as any underwater venture will reveal.

THE PARROT
THAT DOESN'T FLY

ALLOW ME TO INTRODUCE another evolutionary and adaptive marvel: the parrot that doesn't fly! The kakapo—which means "night parrot" in Maori—is the heaviest parrot in the world; it can weigh up to four kilos and has short wings and feathers that keep it grounded. But birds haven't always flown. In all likelihood, feathers didn't develop in order to enable flight so much as to facilitate individual distinctness and communication. In this regard, the kakapo isn't an anomaly; it's a living reminder of extinct birds that never flew in the first place.

The kakapo doesn't fly; it walks. And thanks to its steely claws, it has no difficulty climbing trees. Yet survival proves difficult, despite a life span of ninety years and great skill in the art of seduction. What is this skill? Males make a real impression by "booming." Picture

"Birds are like love— always there. All species vanish, but not birds. Like love."

—Marguerite Duras

the kakapo digging a hole in the ground, inflating a sac in its chest, flapping its wings, and bringing forth inimitable screams—plus a mighty, exploding sound to attract the ladies … and it works! The basin Mr. Kakapo has made amplifies the noise. To this end, he has removed any twigs that might get in the way and carefully chosen a resonant location—for instance, a spot between rock walls or tree trunks. For about eight hours, he makes one "boom" after another, night after night, for three or four months straight. In the process, he can lose up to half his body weight. Depending on the winds, Madame Kakapo will hear these booms as far as five kilometers away; wherever she may be, the sound reaches her, for her beau makes sure to boom in every direction. Females journey to meet their suitors who will fight to the death, parading around, clicking their beaks, and brandishing outstretched wings.

Unfortunately, the heartwarming part of the spectacle stops there. Booming attracts predators, too. Plus, predators come looking because of another of Mr. Kakapo's charms: a moldy smell. This poor parrot is threatened by extinction. Hunted by humans and preyed upon by introduced species (e.g., cats, rats, and dogs)—to say nothing of the effects of deforestation, urbanization, and collecting—the kakapo has all but disappeared.

But sometimes, even human beings prove capable of greatness. In 2012, the fifty remaining kakapos were relocated to three islands without any predators and provided with a constant source of food. It's no easy matter: the kakapo has a very low reproduction rate, and the sex of chicks depends on what the mother eats. If her diet's highly caloric, she has more male offspring. If it's overly rich in proteins, she'll get full too fast—and fail to eat enough calcium to make shells. At first, the kakapos were given fare that was too rich and plentiful, so too many males were born. The wrong balance means trouble. We need to get the diet right. Plus, a single island can't host more than a hundred individuals. The situation presents a challenge for birds and humans alike; together, they're struggling to find solutions and survive.

Once upon a time, there was a parrot that
didn't fly; it built nests and called females
with all the force it could muster; it ate what
it could, clinging to trees and to life itself.

FEEDING,
BY HOOK
OR BY CROOK

THE MOST SKILLFUL BIRD
IN THE SKY

. . . .

ANNA'S HUMMINGBIRD
Calypte anna
Size: 10–12 cm

. . . .

ANNA'S HUMMINGBIRD was named in honor of Anna Masséna, Duchess of Rivoli, an avid bird collector. These tiny creatures, which belong to the family of the smallest birds in the world, consume the nectar of flowers by means of their long, lithe tongue. That said, they're also quite good at defending their territory and are predators able to capture insects in flight—or tear meals from spider webs.

But this hummingbird doesn't perform its greatest feat for food. It really shines in the art of seduction—a performance it stages over and over again! To please his lady love, Anna's hummingbird puts on an amazing show. The male executes a dizzying descent from the heights, a nuptial plunge. The female cannot fail to be charmed as her suitor accelerates at an incredible rate and vibrates his tail feathers to produce sounds much stronger than his voice could achieve. He starts his dive by beating his wings, then shoots down at a speed that can reach about four hundred times his body length per second: some 98 kilometers an hour. This tour de force has earned these hummingbirds the title of the fastest vertebrates in the world—twice as fast as peregrine falcons and fighter planes!

"Poetry is the journal of a sea animal living on land, wanting to fly in the air."

—Carl Sandburg

. . . .

At the end of their descent, they fan out their wings to head back up, accumulating a power of acceleration almost nine times the force of gravity. At a speed like that, it takes exceptional brain and motor skills to avoid obstacles, gauge the height of objects, and steer clear of trees and plants.

Since we're on the subject of records: these diminutive birds with extraordinary aerodynamic properties pull off stunts to make human engineers green with envy. They hover in place, fly backward or upside down, plunge, and change speed and direction even while spinning around. All to get to hard-to-reach flowers. The smallest bird in the world can fly

any-which-way! Because this takes an enormous amount of energy and unparalleled maneuverability, the hummingbird sets the standard for researchers studying flight mechanics. And I haven't even mentioned its fascinating physiological adaptations—for instance, its rate of consuming oxygen (to feed its muscles, in particular), which is the highest of all vertebrates. The hummingbird is a small creature with a big heart: in flight, it beats twelve hundred times per minute!

Finally, nature has endowed Anna's hummingbird with one of its greatest splendors: iridescence. The plumage displays varying chromatism in keeping with the incidence of light. When he turns his head, the male's throat shifts from red to emerald green, to fuchsia. It's always something new! The phenomenon occurs because light penetrates two fine layers of filaments with different refractive indices. Its path and diffraction vary according to the thickness of the plumage and the volume of air between feathers. Think of how soap bubbles glisten with thin, ever-changing rainbows. Iridescent colors are not only spectacular, they also play a key role in hummingbirds' communication and camouflage. What's more, the intensity of a bird's coloration indicates its state of health.

Color doesn't lie! Iridescence provides a trustworthy sign for females, who are always looking for the optimal mate. Undoubtedly, this feature represents a real triumph for hummingbirds in evolutionary terms.

Anna's hummingbird: a creature that, over and over again, by virtue of its tininess as much as its unbelievable grandeur, inspires humility, affection, and admiration.

A COLORFUL PREDATOR

THE PANTHER CHAMELEON
Furcifer pardalis
Size: 20–35 cm (females), 40–55 cm (males)

ONE OF THE BIGGEST chameleons in the world, the panther chameleon lives in the tropical forests of Madagascar and on the island of Réunion, where people have nicknamed it "sleepy" (*endormi*). This creature has adapted to become a formidable predator. It boasts veritable pincers, a protractile and prehensile tongue, eyes that move in different directions, and, most strikingly, skin that changes color. Having also had a small veiled chameleon (*Chamaeleo calyptratus*) hanging onto my finger, I can assure you that these beautiful animals never fail to impress.

An excellent arboreal predator, the panther chameleon hunts by stealth during the day. Appearances might suggest it isn't very good at catching its prey. Needless to say, that's not true. These chameleons stalk slowly. They need a firm grip to move and stabilize themselves before striking, and their feet have adapted to perform this function. Each foot has five toes: two on one side, and three on the other. The design is perfect for climbing and balancing, even when branches are thin. Chameleons freeze in position and wait. Here, of course, their eyes come into play, and what eyes they are! They operate independently and allow the chameleon to look in opposite directions simultaneously. This feature represents a fundamental advantage for a creature that moves slowly and is sometimes not just the hunter but also the hunted. The chameleon can detect prey from any vantage point; whenever the intended meal moves, each of the eyes follows separately to locate it again. When it's time to strike, both eyes synchronize and aim for the target. At this point, another incredible mechanism enters the picture.

> *"It's the equivalent of a human eating a twenty-five-pound hamburger, and having to transport that burger to your mouth using only your tongue."*
> —Kiisa Nishikawa

The panther chameleon's telescoping tongue, which can exceed twice the length of its body, shoots out to seize prey (up to a third of the chameleon's own mass) and bring it back to the mouth. The whole process takes about half a second. By this means, the chameleon can catch not just insects, but even small birds and rodents. The species isn't all slow, either. Prey is captured in the blink of an eye—or, more precisely, 20 milliseconds. This means that the chameleon's tongue reaches a speed of 100 kilometers per hour in less than a hundredth of a second! Before the attack, it's drawn tight, like a catapult; then, the stored energy is released and the tongue blasts forward. And that's not all. The tongue needs to hold prey tight. Yet again, a marvelous evolutionary adaptation comes into play: the tip secretes a dense mucus that's extremely sticky. The substance is four hundred times more viscous than human saliva.

With its claws, eyes, and tongue, the chameleon is an extraordinary beast. To top it all off, its looks are fantastic. The skin of the male panther chameleon displays different colors that reflect the surroundings. The chromatic diversity and intensity are incredible! The chameleon can be green and turquoise blue, but also white and red—or even bright pink and variegated. The same individual may also switch between colors—say, from blue, green, and red to yellow, orange, and white under stressful conditions or during the breeding season. This occurs thanks to nanocrystals located in the epidermis. The male panther chameleon's skin has two thick layers of iridescent cells that, in addition to providing thermal protection, allow it to adopt a wide range of colors for purposes of camouflage, to intimidate other males, or to seduce females.

Chameleons, whether 55-centimeter giants or dwarfs just 2 centimeters long (like the smallest chameleon in the world, Brookesia micra*), possess an improbable number of adaptations for self-defense, motion, detecting prey, and hunting. These animals are a miracle of evolutionary engineering; both for our continued amazement and in hope of explaining their many mysteries, we must protect them at any cost.*

THE DIABOLICAL ANGEL

THE SEA ANGEL
Clione limacina
Size: 2–4 cm

BEHOLD THE SEA ANGEL—also known as the "naked sea butterfly" or "common clione." Don't confuse it with another "angel": *squatina* (the so-called angel shark). *Clione* comes from *Clio* ("to celebrate" in Greek), the muse of epic poetry who praises the glory of cities and men. This gastropod, down at a depth of more than 500 meters, maintains a vertical equilibrium position, pointed toward the surface. Its head is equipped with two pairs of short tentacles in front of its eyes. The sea angel has no shell or gills; the exchange of respiratory gases occurs through its gelatinous skin. Like other mollusks—snails and oysters, for instance—these small animals are hermaphroditic. They reproduce sexually, but when they meet a partner, they can assume the role of either male or female.

It's an important source of food for cetaceans, "flying" delicately and gracefully in the water. That said, one shouldn't trust appearances. Behind a harmless exterior lurks a fearsome predator. The sea angel consumes zooplankton like the gastropod *Limacina helicina*, its favorite meal. Thanks to lateral fins, it swims very fast. Once the sea angel has detected prey, it opens its mouth in a flash—10 to 20 milliseconds—then deploys six tentacles that are otherwise retracted into its head (as in the illustration). These appendages grip the prey and secrete a viscous material that's especially sticky for shells. Then it rotates the victim so the soft parts are exposed. Next, the hunter opens two sacs, each containing some thirty hooks. These chitin hooks, which are tough but flexible, shoot out and pierce the prey, allowing the sea angel to extract all that's edible. A raspy tongue—the *radula*—licks the victim to death. In half an hour, the prey's shell has been cleaned out and its contents consumed. So is this angel really a devil?

This tiny marine gastropod seems to fly, like an angel, in the glacial, dark, mysterious, and abyssal waters of the Arctic Ocean.

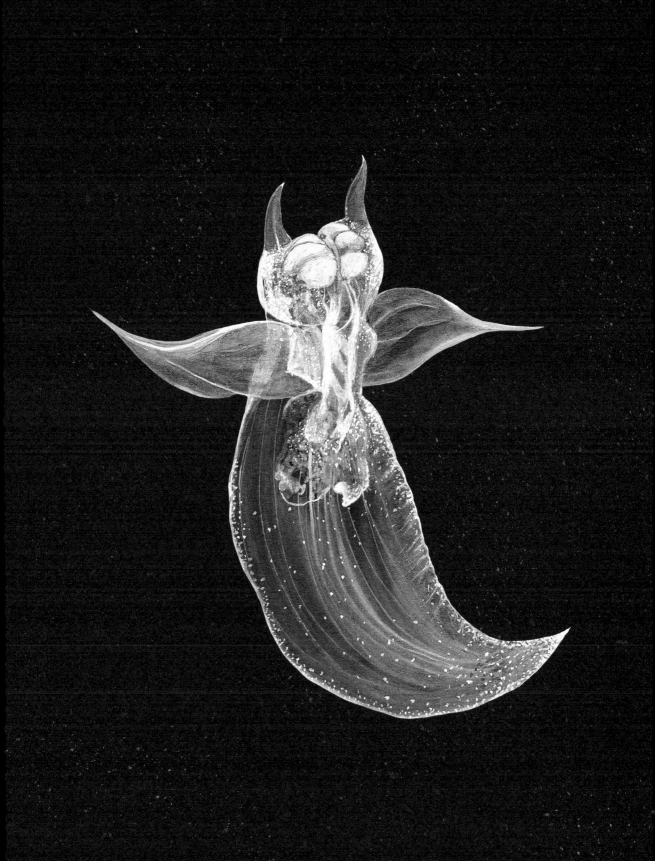

Another feature lending the creature an air of poetic mystery is a translucent body that reveals its inner organs. This magnificent little mollusk seems to have come from another time and place. It moves rhythmically, with natatory organs shaped like wings that oscillate and propel it. Interestingly, this locomotion correlates with specific cardiac activity. Sea angels take regular breaks when traveling. After a few minutes of movement, they stop abruptly. During periods of "rest," its heart stops beating. The same connection has been observed, in reverse, when the animal is defending itself or hunting; at these times, its heart rate increases. The cardiac physiology of this small mollusk harmonizes perfectly with its activities.

"*The shore was strewn with mollusks, little mussels, limpets, smooth bucards in the shape of a heart, and particularly some clios, with oblong membranous bodies. … I also saw myriads of northern clios, one and a quarter inches long, of which a whale would swallow a whole world at a mouthful. These charming pteropods, perfect sea-butterflies, animated the waters on the skirts of the shore.*"

—*Pierre Aronnax, Professor of Natural History at the Museum of Paris, in Jules Verne,* Twenty Thousand Leagues under the Sea

THE BOXER OF THE SEA

THE PEACOCK MANTIS SHRIMP
Odontodactylus scyllarus
Size: 3–18 cm

THE PEACOCK MANTIS SHRIMP is a crustacean inhabiting deep-sea corals, on the rocky or sandy floor, at depths of up to 70 meters. It either dwells in a burrow of its own or settles in one dug by another animal. Days are spent in hiding, but at night it hunts for mollusks, other crustaceans, and fish. The peacock mantis shrimp is a marvel of nature. This magnificent creature is often considered the most aggressive ocean predator; the slightest incursion onto its territory meets with swift reprisal. And to think I thought crayfish were aggressive (having been pinched on more than one occasion)… well, the peacock mantis shrimp is worse! Even divers make sure to avoid it. It's also known as the "painted mantis shrimp" or "clown mantis shrimp" because of its bright colors and murderous limbs. The name sounds funny, but it's no joke.

This animal's limbs are veritable engines of war—even more fearsome than those of the earthbound mantis. In fact, they're so light and tough that scientists study them in the hope of creating new protective materials for human use. The appendages serve any number of purposes: by turns, the peacock mantis shrimp uses them to stun or crack open prey, dig burrows, defend itself from predators, and fight against its peers. The firm and round ends, though just a few millimeters in size, permit the diminutive killer to break open shells in a series of blows; then, it consumes the victim's flesh. The peacock mantis shrimp is a real boxer. Its

> *"Anyone who claims that nothing surprises him has never viewed a lobster face-to-face."*
> —Auguste Villiers de L'Isle-Adam

hammering is incredibly rapid, precise, and brutal: a speed of 100 kilometers per hour, landing with a force up to 1,501 newtons! The motion is among the fastest in the natural world—as if the little beast were firing bullets with a force of impact thousands of times its body weight! The attack occurs so rapidly, it's invisible to the human eye—and even

more impressive in light of the water resistance. The blows come so hard and fast that even when it misses, the peacock mantis shrimp manages to kill its prey. If we could throw a ball at the same speed, relative to our own weight and size, it would wind up in orbit!

These arms aren't the animal's only resource when defending itself and hunting. The peacock mantis shrimp also has extraordinary vision. Its optical organs comprise thousands of facets and a large fovea—one of the most elaborate visual systems in the animal world. It's a real marvel of nanotechnology! Each eye functions independently, takes in a full 360-degree range stereoscopically, and detects objects with three different parts of the same organ to assess depth and distance—and aim punches. Their retinas also have the largest number of photoreceptors for color in the animal world: twelve in each eye. Most animals have only two to four photoreceptors; human beings have three (red, blue, and green).

This unique feature allows the peacock mantis shrimp to make out colors rapidly, which is especially useful for an animal that's vibrant itself and lives in variegated coral reefs. It's a vital resource for fighting, fleeing, communicating, and reproducing. Indeed, the peacock mantis shrimp can even register the polarization of light (i.e., the oscillation direction of electric fields), multispectral images across the electromagnetic spectrum, and ultraviolet rays—a real boon for locating food. Its incredible eyes scan the surroundings to identify different types of coral, follow transparent prey, and detect approaching predators like barracuda. What's more, they enable the peacock mantis shrimp to tell the tidal phases of the moon, know when females are fertile, and sense a partner's fluorescence. Without question, this crustacean's eyes are the most complex in the animal world—an indication of the boundless power of evolution.

The compound eyes of the peacock mantis shrimp perceive polarized light—which can reveal whether tissue is cancerous or healthy. Thus, researchers are trying to make cameras with this same feature in order to detect disease. Just think: a crustacean in the service of preventive medicine …

A FAR-SIGHTED BIRD

THE TAIWAN BLUE MAGPIE
Urocissa caerulea
Size: 63–68 cm

VOTED THE COUNTRY'S national bird in 2008, the Taiwan blue magpie, or "long-tailed mountain lady," lives in forests. Like other members of the family *Corvidae*, it makes hoarse and noisy sounds. Sometimes it ventures into urban areas looking for food. These magpies move in small groups of at least six individuals, and they aren't frightened by humans—though they are cautious and try to avoid them as much as possible.

The Taiwan blue magpie enjoys a highly varied diet: fruit like figs, papayas, and berries, seeds, and plants, as well as insects, small invertebrates, rodents, and snakes. It's an omnivore and sometimes a scavenger, too. Individual members of a group scout out meals by surveying the forest canopy, flying from one tree to the next; the species is excellent at navigation. In addition, Taiwan blue magpies make sure to hold supplies in reserve, in case of scarcity. They usually hide surplus food in the ground and cover it with leaves; sometimes they also hide it up in trees. These stashes are serious business.

Needless to say, finding and storing extra sustenance requires good memory. Taiwan blue magpies aren't birdbrains! Their brains may be small, but they're also well made. It's interesting to note a feature shared by certain birds and certain human beings. Birds that cache food have a bigger hippocampus (the part of the brain responsible for memorization and navigation) than birds that don't. The organ plays a fundamental role in locating hiding places and being able to remember where to find them. Indeed, researchers have observed a peak of neural recruitment to the hippocampus when a bird is recovering hidden seeds. The same rule holds for people. A study of London taxi drivers has demonstrated that the hippocampus grows in this line of work; the more time drivers spend studying city maps, the more neurons are recruited to the organ. People who drive cabs and birds that stock food—a united front, it might seem. But in actual fact, it's not that simple. Some caching birds don't have a large hippocampus and probably use other neural substrates—which still points to fascinating cerebral flexibility.

The Taiwan blue magpie doesn't just eat and store food, of course. It also breeds! These birds are monogamous, and a spirit of cooperation prevails between partners. Females incubate the three to eight eggs on their own, but both parents participate in constructing the nest and raising young. The female also defends the nest quite aggressively, rushing headlong at intruders until they leave; these mommies are so concerned about their babies that attacks on human beings occur fairly regularly.

Finally, the Taiwan blue magpie, like other corvids (e.g., crows, ravens, and rooks, *Corvus sp.*), is known for its cognitive abilities. These species can use tools. For example, they find branches with small irregularities to "fish" larvae out of holes in trees, and they cast pebbles into water to raise its level, so they can gather floating seeds more easily. It seems these species have a sophisticated understanding of objects' material properties. Indeed, some corvids can conduct operations on a massive scale. In North America, Clark's nutcracker (*Nucifraga columbiana*) stocks up reserves during the autumn by hiding four or five seeds at a time in almost ten thousand caches. In the winter, it moves to lower altitudes, where life's easier. Some six months later, it goes back to recover its provisions—and manages to find about three thousand of the hiding places. (By way of comparison: I have trouble finding my keys in the morning … and not in the mountains, either—just in my apartment!) Now, the diminutive bird can leisurely enjoy about one-third of its stash, in spite of all the damage done by snow and other animals in the interim. Pretty impressive!

*"The bird's body is made out of
the air that surrounds it,
and its life is made up of the
motion that transports it."*
—*Gaston Bachelard*

SEDUCING
TO LIVE

A MUSICIAN THE LADIES LOVE

....

THE GREAT BLACK COCKATOO
Probosciger aterrimus
Size: 60–70 cm

....

THE SEDUCTIVE POWER OF MUSIC … human beings aren't the only ones to use this stratagem. A case in point is the sublime great black cockatoo. Its attire is entirely black—beak and legs included—except for a red spot on each cheek; this featherless area can change color if the bird is feeling stressed or sick. And unlike the kakapo, this parrot (which weighs a kilo at most) can fly! A mighty beak cracks open nuts with ease, and the tongue, which is almost as dextrous as those of macaws I've studied, is quite good at extracting food. The great black cockatoo doesn't have a restricted diet; it consumes seeds, plants, berries, insects, and even larvae. Birds and whales are known for their lovely songs. But this cockatoo doesn't just sing; the male also plays an instrument.

To be more precise, the great black cockatoo of Australia is a drummer! To attract the ladies and win their hearts, the male finds a stick about 20 centimeters long. Grabbing it with his feet, he then starts hitting tree trunks. The sound can be heard over a hundred meters away. But why is this music, and not just noise? Because these birds strike practically perfect cadences, and for an extended period. Human beings do the same when they play percussion: they keep a steady beat over time. The situation is even more interesting because each great black cockatoo plays in a style of his own, at his own pace. Some always go fast, while others are slow—or prefer to play in a somewhat fanciful manner. Personal signatures help other parrots—especially females—know who's making the music. It's a real ritual of individualized seduction.

> *"Poets are birds: any sound will make them sing."*
> —*François-René*
>   *de Chateaubriand*
>
> ....

When the musical ploy works, mating ensues. The female lays an egg up high, in the hollow of a tree. And the male isn't just a good musician. He's also a conscientious father; together, both partners incubate the egg for thirty days. Once it's hatched, the chick will

remain in the nest for a little over three months. The great black cockatoo can live for a hundred years—or more! When circumstances are favorable, anyway. Unfortunately, a low reproduction rate and the loss of habitat through mining have placed these birds at risk.

Parrots have fascinated people for centuries. Their colors, intelligence, and, in some instances, ability to speak have captivated us ever since explorers, corsairs, and pirates first brought them to Europe.

AN ARCHITECT AND DESIGNER

THE SATIN BOWERBIRD
Ptilonorhynchus violaceus
Size: 30–35 cm

BIRDS HAVE AN INCOMPARABLE resource for seduction: color. Male birds-of-paradise display extraordinarily beautiful adornments to dazzle the ladies. But those who can't boast finery need to be creative. Satin bowerbirds (*Ptilonorhynchus violaceus*) don't have the option of parading iridescent plumage before potential mates, so they enlist a different strategy. To please the opposite sex, they make elaborate structures. These constructions, which are highly varied and intricate, draw on whatever's available in eucalyptus forests, the abundant flowers of mangroves, and the offerings of urban parks and gardens—including plastic bottle caps!

Bowerbirds spend the winter in groups of about fifty. During spring and summer, they're basically solitary; males become quite territorial and protect their space by strutting about imperiously. Their most striking activity occurs when Mr. Bowerbird starts building. It's not just any building, either, but a veritable honeymoon suite. He gathers everything available to provide an optimal setting for wooing his beloved. No effort is spared. For weeks on end, he adds to the design. It starts with a platform of twigs in an open area that's been chosen carefully. Then, he arranges two rows of branches to make a narrow tunnel about 90 centimeters in length, sometimes with an arch at one end. The branches, which are usually between 30 and 35 centimeters long, form a vault. Once the basic work is done, he decorates the entrance and makes a nuptial courtyard. It includes pebbles, feathers, berries, flowers of every color, leaves, shells, and anything else he's managed to find—desiccated insects, jewelry, even bones! Often, he'll make sure the female sees the splendor from a particular angle. It's a surprise!

After laying the foundations and finishing the floor, he decorates the walls. Sometimes, he even uses a brush! Mr. Bowerbird grabs a piece of bark with his beak and applies broad strokes of a compound made from berry juice, charcoal dust (from bush fires), and saliva. His attention to detail is great, and he often chooses blue berries; this color, which is close

to the hue of his own plumage, evidently attracts the ladies even more. Some males repaint the walls every day.

How can the female fail to be seduced after he's taken such pains? Well, the male's work isn't done: Mr. Bowerbird has to get the attention of his lady love for the "chick magnet" to work its magic. He launches into a spectacular parade: hopping around, fanning his wings, shaking his tail, stretching his neck out, and emitting rhythmical sounds. When, at long last, she deigns to notice him, he offers a gift and invites her in.

She might still take weeks to decide; Ms. Bowerbird isn't easy to please. Once he's won her over, they mate under the boughs; then she flies off from this colorful setting to build a nest for the eggs she will lay in one or two weeks. The female defends her eggs and young against predators by herself. In due time, these chicks will learn how to build marvels of their own. In the meanwhile, the male maintains his property and keeps up the parades to attract other ladies. He can do so for quite some time—up to fifteen years, in fact. When a dominant male dies, his peers fight for possession of his bower. These gentlemen know the value of good real estate!

"It's the priest that makes the church, and the king the castle. And the winter? The North wind. And the nest? What, then, if not love?"
—*Victor Hugo*

When not busy pitching woo, the satin bowerbird feeds on fruits, flowers, seeds, nectar, insects, and invertebrates. These birds forage a great deal on the ground and can cause real damage when they get into orchards. They've paid the price, alas, and have been all but wiped out in the area surrounding Melbourne. Fortunately, the satin bowerbird has more than one trick up its sleeve. Because they have such a varied diet, bowerbirds are also great mimics. They produce an impressive array of sounds; whistles, buzzing, and hisses all form part of the creative palette. Indeed, bowerbirds enjoy full poetic license: they imitate other birds (including predators), meowing cats, barking dogs—and even noises made by human activity.

THE UNDERWATER SCULPTOR

....

THE PUFFERFISH
Torquigener sp.
Size: 10–20 cm

....

PUFFERFISH ARE FOUND in waters the world over. They inhabit coral reefs, where they feed on algae, mollusks, sponges, corals, crustaceans, and echinoderms. Thanks to a powerful jaw equipped with four teeth, they lead an opportunistic, omnivorous existence. The same holds for their impressive cousin, the diodon (*Diodon sp.*), or porcupinefish, which has mighty spines (see the illustration) and can also inflate to intimidate predators. The pufferfish itself has neither spine nor scales. Pretty unusual for a fish, you say? Well, the pufferfish holds even more surprises in store! Above all, it's a wondrously creative architect and engineer; the diodon, however imposing it might otherwise be, can't say the same.

Picture the depths, 30 meters down, off the coast of Amami-Oshima, an island in southern Japan. Here, the pufferfish does something that puzzled researchers for years. This fish, which is about 12 centimeters long, carves magnificent, rose-shaped labyrinths on the ocean floor—circular patterns of remarkable symmetry, which can measure up to two meters in diameter (some fifteen times the artist's own size). It takes seven to nine days to finish the job. So why does the male pufferfish take all this trouble? To win a mate, of course! The result is an underwater fresco in relief, crafted by an artist-engineer who digs furrows in the sand—small, parallel dunes—with his body and fins. And the pufferfish doesn't stop there. His aesthetic sensibility also finds expression in the shells he distributes over the work he's made. Critics will have the last word, I suppose, but it looks like he's offering a real piece of art. Yet another remarkable fact: the more complex the design, the greater the male's success with the fairer sex. Once the female has finished her inspection, the couple mates. Now, the ridges keep the eggs that are incubating at the middle of the circle from drifting away. In addition to looking good, then, the design protects the eggs; what's more, the scattered shells provide food for hatchlings during the first hours of life. However one defines it, art is a part of nature.

Apart from this incredible artistic gift, the pufferfish possesses two major traits—which it shares with its cousin, the diodon—for protecting itself and warding off aggressors. First, it can swell up in a matter of seconds by gathering air or water in its esophagus; the spherical shape the fish assumes is intimidating and difficult to bite. Second, it produces a potent neurotoxin—tetrodotoxin—which is twelve hundred times more poisonous than cyanide! This substance, synthesized by bacteria in algae that the fish consumes, paralyzes the attacker's muscles and can cause death by respiratory arrest. It's quite the trick: pufferfish and diodons themselves aren't affected by it, which allows them to build up tetrodotoxin in almost all their organs (especially the viscera and ovaries). A unique feature of diodons is the intense concentration of poison at the tip of their spines; during the mating season, females can administer a particularly strong dose. This toxin holds a good many predators at bay—to say nothing of cooks and diners!

Indeed, tetrodotoxin is present in a celebrated pufferfish, fugu (*Takifugu rubripes*). In Japan, Polynesia, and Taiwan, the animal is considered a delicacy. Long ago, the Emperor of Japan and samurai were forbidden to eat it, and the law still holds for the Emperor. Why is this? Well, if the fish isn't prepared properly, it can cause death in just a few hours; for someone who weighs 75 kilos, all it takes is 25 milligrams! There's

> *"Most people think I count fish but I don't, I look at them. I look at their souls. And read their dreams. Then I let them into my dreams. People think that fish are stupid. But I was always sure that they weren't because they know when to be quiet, and it's people that are stupid. They pretend they know everything and don't need to think."*
> —*Emir Kusturica,* Arizona Dream

no antidote. This deadly quality has made the fish a star in literature and film, serving the nefarious purposes of assorted killers—including a food critic in an episode of *Columbo*.

All fiction aside, tetrodotoxin eliminates predators effectively, but that's not all it does. Here, the story gets pretty interesting: in small doses, the poison's a hallucinogen. There's a video, about thirty minutes long, showing dolphins playing with a pufferfish off the coast of Mozambique. Feeling threatened (and for good reason), the pufferfish secretes its neurotoxin, and the dolphins seem to become hypnotized, in a trance! Eventually, the fish escapes. Good for him: consumption has grown to the point that the population of some pufferfish has fallen by 99.99 percent in just forty years.

A TINY SEDUCER

THE PEACOCK SPIDER
Maratus volans
Size: 3–5 mm

WHEN THEY'RE NOT SCARING PEOPLE, spiders are celebrated for their webs: masterpieces of architecture and technique, and fearsome traps, too. They aren't as well known for their seductive talents, but these creatures can be quite charming! In particular, there's a jumping spider that has no equal when it comes to putting on a show: the tiny peacock spider (*Maratus volans*), a little wonder with excellent vision, which has really been known to us only since the early 2000s.

Thanks to their scales—whose structure is more complex than that of butterflies— males display a rich array of colors. It's a unique design that serves to entice females. Not only does the male peacock spider boast red, yellow, and white pigmentation; the scales also constitute an optical system, filtering the sunlight and reflecting striking shades of blue and purple. And that's not all. The scales have an extremely sophisticated architecture. They consist of two chitinous layers with ridges less than one micrometer apart; the inner surface of these ridges contains a network of parallel filaments with just one-tenth of a micrometer in between. This is how the play of light produces the blue color. Though tiny, this spider is complex and sublime.

"Spiders are thought to be dangerous, hairy, and gloomy. Such ideas, which are basically wrong, transmit fear and disgust. Part of my work is to see that spiders are treated fairly."
—Christine Rollard

So small, but ... so winsome! For starters, on his opisthosoma—the posterior region of the body (abdomen)—Mr. Spider has a sublime pattern in blue, red, and yellow or orange. What's more, he knows how to use it. How, you ask? By strutting his stuff. Like peacocks, these little guys make eye contact with potential mates before rising up and exposing their abdomen to let their colors shine. And there's more. The male shows off to the female from the best

possible position—in keeping with how the light falls—so the brightness and colors will vary.

Sometimes, he'll perform an abrupt jump. In fact, there's quite a lot at stake. The female is a little bigger than he is and capable of eating her paramour for lunch! He just might be dancing for his life. The male has every incentive to succeed, and he spares no effort. Depending on the female's reactions, he'll start again and again, parading for seconds or minutes at a time. To maximize his chances, he produces vibrations and visual signals by positioning his limbs in different ways—near or far apart, depending on the species. The spectacle also involves shaking his iridescent abdomen, waving around pedipalps (appendages around the mouth), displaying the black and white bristles on his third pair of legs, hopping around, and so forth. If he pulls it off, the elaborate dance makes the female's head spin!

By this means, the male shows the female he's ready to breed. The process shows sexual selection at work. Males are in competition with each other, and the decision belongs to the female. Each one puts on a show of his own, with various types of movement, vibratory signals, speeds, levels of chromatic intensity, and so on. So the female's choice is based on specific signs that suitors provide. By using high-speed cameras to film and quantify these performances, researchers have determined that success depends on a combination of visual and vibratory cues. Evidently, the female makes her pick in keeping with the complexity of the male's parade; she might be gauging his power, dynamism, and state of health. That said, the elaborate display might also serve to lull the female into a hypnotized state—to diminish her natural aggressiveness and lower the chances for real trouble!

This tiny jumping spider is as beautiful as its dance. After mating, the female lays her eggs; they hatch after two weeks. The young first stay with their mother; after another fifteen days, they're off. Once they've reached adulthood, male offspring will bring forth rainbows of their own, the smallest in the world.

THE FLYING TURTLE

THE PIG-NOSED TURTLE
Carettochelys insculpta
Size: 55–75 cm

THE PIG-NOSED TURTLE: what a strange and unique reptile. As it glides along, graceful and calm, it seems to be flying. It's the biggest freshwater turtle alive, the only survivor of a line that was widespread 40 million years ago. It's also the only turtle equipped with a prominent snout for flushing out larvae, crustaceans, fish, snails, worms, algae, leaves, or fruit that has fallen into the water. This creature is an opportunistic eater, so it feeds on carrion when necessary.

The pig-nosed turtle takes advantage of the wide range of food available. But eating isn't enough to ensure survival, which is also a matter of perpetuating the species. These animals reproduce only every two years. The pig-nosed turtle takes care to optimize the conditions for its eggs and hatchlings by means of two remarkable strategies. During the dry season, the female emerges from the water to lay her eggs. In the sand, she digs a nest about fifty centimeters deep and deposits a clutch of about ten (although it can include up to forty). If the nest is very hot (over 32°C, or 90°F), the young will be female, whereas cooler conditions produce males; if the temperature is somewhere in between, hatchlings of both sexes are born. Like sea turtles, pig-nosed turtles lay their eggs on the beach. As soon as they hatch, the young of the former are exposed to terrestrial predators; in consequence, very few of them ever reach the water. But pig-nosed turtles have found a way of avoiding this problem: their eggs don't hatch in the sand. What miracle makes this possible? Mother Nature guides them … the eggs wait until the water level rises! At the end of the dry season, when water immerses the nest, the young emerge. Just like that, they're in their element. It also helps that their shells are equipped with

"Don't mock the turtle because it's humble; it might be your guide tomorrow."
—Ovambo proverb, southern Africa (Namibia and Angola)

"teeth" which make them look like a circular sawblade; this represents another adaptive resource for self-protection, giving the young turtles a chance to grow up.

The stakes are high. These turtles face danger not only from predators such as crocodiles and lizards, but also, more indirectly, from herds of buffalo (which can trample on nests, eggs, and hatchlings). Plus, invasive species such as the cane toad eat their eggs. Nevertheless, the biggest threat is surely humankind. Even though these turtles hold an important position in the art and legends of native peoples, their habitat is being destroyed by mining and agriculture. They're also caught and sold as pets—or consumed for their meat and eggs. In forty years, the pig-nosed turtle population has declined by more than 50 percent. The scale of destruction is vast: one to two million eggs are collected for international trade every year. If drastic measures aren't taken to rein in commerce, the animal will soon be threatened by extinction.

Unfortunately, pig-nosed turtles aren't the exception: two-thirds of all turtles are threatened or endangered, mainly because they're taken from the wild and sold. Unless things improve, this species will grow rarer and rarer. All the same, there is some hope: the pig-nosed turtle is now protected by the Washington Convention, and breeding programs have begun, especially in Australia.

Once upon a time, there was a turtle that flew … in the water.
Without knowing the implications, modestly and discreetly,
it explored the depths. Its activities redounded to the benefit of
human archaeology. How? By prompting engineers to design
a small, submarine robot—a U-CAT—with four independently
motorized fins; because these fins don't disturb the water
like a propeller, the craft can navigate through shipwrecks
easily. Nature is the source of all inspiration.

SOLITARY REPRODUCTION

THE GREAT HAMMERHEAD
Sphyrna mokarran
Size: 5–6 m

AS I SEE IT, the great hammerhead—a beautiful and imposing creature that all but defies belief—rules the seas. Not only does it seem to come from another time; with those broad, flat extensions of its head and pronounced dorsal fin, it's also one of the most enigmatic inhabitants of the depths. Despite its commanding stature and weight (a ton), it moves gracefully through pelagic expanses (the high sea) and in coastal regions. Most intriguing of all, however, isn't its elegance so much as how it reproduces.

The way great hammerhead sharks breed is still veiled in mystery because observations are so hard to conduct. It seems that males parade around females, seeking to attract the best partners they can, and when night falls, couples head off to mate in private. Males who haven't found a match try their luck again the next day. There's nothing surprising about that. But listen to what happened in 2001 at Henry Doorly Zoo in Omaha, Nebraska. Here, three female shovelhead sharks (*Sphyrna tiburo*), a species closely related to the great hammerhead, occupied the same tank. To universal amazement, one of them gave birth. Was it a miracle? Not quite: researchers observed that females have the ability to store males' sperm for later use. Yet the maximum storage period is five months, and the shark in question had last encountered males three years earlier. So another possibility had to be considered: parthenogenesis, asexual reproduction permitting females to generate young on their own. But this phenomenon had never been observed in sharks. Genetic tests were conducted and, indeed, the newborn had no trace of a paternal chromosome. It's the first case of parthenogenesis observed in a shark.

"The ocean gets its saltiness from the tears of misunderstood sharks."
—*Hypnotide,* Landlocked

However they reproduce, hammerheads give live birth, following a gestation period that lasts for nine to twelve months. The female carries ten to forty embryos; the young sharks measure about sixty

centimeters when born. Incidentally, researchers have discovered a veritable nursery for great hammerhead sharks in waters protected by the mangrove forest and reefs of the Galapagos. For millions of years, female hammerheads have been going there to give birth; here, their babies are safe from predators and find all the food they need.

In addition to possessing this extraordinary mode of reproduction, the great hammerhead shark deserves recognition as a predator. It feeds on fish, other sharks, crustaceans, cephalopods, and the like, and is immune to the venom of leopard rays—its favorite prey. Great hammerheads have a highly evolved sense of taste, smell, and hearing. They can track down scents at great distances and detect the slightest change in water pressure. Not to mention that flat head, which gives it excellent stereoscopic vision—and in deep, dark waters, at that. Its skin is also covered with scales, or denticles; they point backward, making it rough and, more importantly, exceptionally hydrodynamic. Finally, the great hammerhead shark is very sensitive to electricity: with a kind of sixth sense, it perceives electrical fields, including the muscular contractions of prey hidden in the sand.

These qualities contribute to a defining feature: an excellent capacity for navigation. Great hammerheads move and migrate in huge groups; hundreds gather at regular intervals to feed or reproduce. In a single year, they can travel thousands of kilometers, following paths even more complex than the itineraries of birds. They do so thanks to a system of dorsal electrosensors connected to receptors located around the snout and head. Thus, in addition to using the position of the sun or the moon, great hammerhead sharks find their way by means of the Earth's electromagnetic field, as well as electrical fields produced by other animals and ocean currents. This internal geomagnetic compass explains their prodigious ability to orient themselves and gather during the breeding season.

Unfortunately, of the more than five hundred species of sharks in the world, one hundred and eighty are endangered—and thirty of these species face extinction, including the great hammerhead shark. On average, 100 million sharks are killed every year—that is, three sharks per second. Why such slaughter? Because of a mythical prejudice, or so their fins can wind up in a soup. Yet the fabled bloodthirstiness of sharks is ridiculous. Every year, only five deadly shark attacks occur—in contrast to thirty deaths caused by dogs, one hundred by jellyfish, six hundred by elephants, and five thousand by scorpions. Is there any hope? Yes, thanks to sanctuaries in Ecuador, French Polynesia, Palau, the Maldives, Honduras, the Bahamas, and Tokelau. Plus, there's another bit of good news: research now suggests that shark fins are actually poisonous for humans to eat!

Having lived on our planet for 450 million years, sharks are as frightening as they are fascinating. They've conquered almost all the oceans, and play a fundamental role there. Yet all the same, extinction rates have reached a crisis level. This magnificent, unloved creature seems fated to disappear, the victim of a silent, slow, and cruel program of genocide—it's simply unbearable.

SELF-CARE,
REGENERATION,
RESISTANCE,
STOPPING TIME

THE MASTER OF SELF-MEDICATION

THE CHIMPANZEE
Pan troglodytes
Size: 1.30–1.70 m

THESE GREAT APES are primates, like us. Sure, they're a little hairier, but they're hominids. Like us, these creatures are amazing. Like us? It's a risky comparison. In many respects, chimpanzees are more accomplished. One talent they possess would benefit us, too: they know how to take care of themselves. Since the 1970s, researchers have known that chimpanzees—especially those in Tanzania or Uganda—use medicinal plants. They consume fruits with antimicrobial properties; sometimes they combine them with other substances to reduce the toxicity. Other chimpanzees eat flowers with antibiotic properties or leaves with antiparasitic ones, which act as laxatives or even induce uterine contractions. Chimpanzees also tear bark off trees and lick the resin to kill internal worms; the compounds, tests in vitro have shown, slow the growth of cancerous cells.

"Endless fascination, endless enjoyment, endless work."
—Jane Goodall

Significantly, practices of self-medication vary between chimpanzee populations. When chimpanzees feel sick, they seek out a particular tree and ingest a few leaves. The bitter leaves contain molecules that are quite effective against plasmodium parasites, which cause malaria. But chimpanzees also consume about ten other kinds of plant to combat these organisms. Thus, in contrast to human beings (who use a small number of substances for warding off malaria), chimpanzees diversify their medical arsenal. What's more, when making their bed for the night, chimpanzees in Uganda do so in areas where there are fewer mosquitos. Do they choose plants based on their potential for repelling pests, or is softness—and resulting comfort—the deciding factor? We'll see.

For some time now, chimpanzees' pharmacopoeia has been the object of study. Indeed, research by Jane Goodall in the 1960s even prompted scientists to reexamine traits thought to be exclusively human. In fact, chimpanzees use an array of tools for different purposes:

branches for digging out termites, honey, or marrow; sticks and stones for cracking nuts; and sharpened pieces of wood for spearing galagos (bushbabies). They even make "shoes" to protect their feet when climbing thorny trunks. Using these tools can be complex and require training; some mothers actively show their young the right way. Pedagogy, then, is another practice we share with chimpanzees. Another exciting finding: techniques differ from one population to the next (e.g., Uganda, Ivory Coast, Guinea). Many writers on the subject don't hesitate to speak of "traditions" and "cultures." This observation raises another set of questions. Do chimpanzees invent? In Tai National Park, Ivory Coast, generations of chimpanzees were known to use branches to break extremely hard nuts (*Dura laboriosa*). One day, a female member of the group, Eureka, employed a stone for the same purpose and continued to do so in the presence of her companions. And then? Other chimpanzees started doing the same. After a few generations, the entire population had switched tools for cracking nuts, from sticks to stones.

Chimpanzees use tools, but they're even better at something else: memorization. The same chimpanzees in Ivory Coast have a geometrical understanding of their territory, which spans twenty-five square kilometers; they move from one spot to another in more or less straight lines. Even with a limited range of vision—thirty meters, at most—they know where to go to find ripe fruit and avoid danger (including rival chimpanzees!). By remembering topographical features of the landscape and picturing abstract space, they can calculate distance and direction, no matter where they are.

And that's not the only proof of chimpanzees' cognitive abilities. In a computer-based test of spatial memory, researchers compared young chimpanzees and university students. The experiment involved clicking numbers one by one, in the right order. At a late stage, the task became more complicated: as soon as subjects clicked the first number, a white square blanked out the other ones; the point was still to press the right series, in order to receive a reward. And the results? The chimpanzees pulled it off 80 percent of the time—that is, twice as often as the students did. From an early age, these animals demonstrate highly developed visual memory; it's almost photographic. This is what enables them to memorize where the best fruit is and determine the best path to take. That is, when they wouldn't rather break our cameras—or throw all kinds of stuff at us.

Traditional healers and villagers use the same parts of plants as chimpanzees, and for similar purposes. Of the 500,000 kinds of plant on the globe, only 10 percent have been studied for their biological and chemical properties. Undoubtedly, the plants that chimpanzees pick represent the medicines of tomorrow.

THE "WATER MONSTER" THAT REGROWS ITS BODY

THE AXOLOTL
Ambystoma mexicanum
Size: 20–30 cm

HOW DO YOU MAKE something new out of something old? Ask an axolotl! The animal the Aztecs called "water monster" belongs to the family of *urodeles*, which includes salamanders and newts. For the most part, it lives in lakes up at an altitude of 2,000 meters; it's also found in volcanic craters filled with water. In the wild, axolotls are usually dark, but some of them just have dark eyes and are otherwise depigmented. These animals can spend their whole lives in the larval state; they're capable of reproduction, but they still breathe through their skin as well as lungs and fern-shaped gills around the head. In other words, the species is neotenic; individuals reach adulthood only when water levels fall dramatically and the weather warms up. Having viewed axolotls up close, I can assure you that they're impressive creatures. But I still haven't mentioned their most remarkable trait.

Lizards, salamanders, crabs, and octopuses can all regrow limbs—or even organs—if they're seriously injured or severed from the body. The axolotl possesses the same ability, and to a fantastic degree. When an organ is damaged or destroyed, a small bump—what's known as a *blastema*—forms. In the course of a few weeks, a new member identical to the one that's gone missing takes shape. The axolotl can regrow a leg, its tail, or an eye in the space of a month, if not sooner. But these creatures can do even better than that: they even rebuild parts of their brain! Moreover, axolotls do very well with transplants, prove extremely resistant to cancer, and produce eggs throughout their entire lives.

This animal is highly studied because of the prospects it opens for regenerative medicine and fertility. It would be incredible to identify the biological markers that trigger regeneration and use this insight to benefit human beings. Now, such projects are no longer utopian. Blastema cells have long been recognized as pluripotent; that is, they can differentiate into any organ. Mistakenly, it was long believed that these cells were embryonic, and that adults no longer produce them. In fact, however, cells "remember" their function—thus, the cells of old muscles can lead to the formation of new muscles. Since

the cells involved in regeneration aren't necessarily embryonic, it might be possible to find them in adult mammals. Of course, the problem is far from being solved, because—for millions of years now—mammals have healed by forming scars. We have no idea how they would react to another form of healing that proceeds by cell proliferation—which is exactly what happens with cancer.

At any rate, no matter what dreams people conceive when contemplating axolotls, we should be worried about their survival in Lake Xochimilco; in just ten years, the population has dropped from six thousand individuals per square kilometer to just one hundred in the same space. Although one axolotl can lay up to 1,500 eggs, and do so four times a year, pollution, fishing, and the exotic pet trade are getting the better of the species. It's in critical danger of extinction and may soon disappear.

Axolotls symbolize an elementary and primitive form of life, present from its very inception. Like the frog in ancient Egypt, they represent the "obscure forces of a world that's still inorganic [...], the spontaneous creation of primordial waters" (Georges Posener).

FOREVER YOUNG

THE IMMORTAL JELLYFISH
Turritopsis nutricula
Size: 5 mm

SMALL IN SIZE but great in accomplishments, this jellyfish is one of the oldest animals on the planet; fossilized remains from 650 years ago have been discovered. It's 98 percent water and has no skeleton or brain. Although the thousand jellyfish species on record include some that prove deadly at the slightest touch and others with twenty-four eyes, first prize still goes to *Turritopsis nutricula*. It's the only animal that can reverse its cells' aging process. That's right: this jellyfish is able to grow younger!

More precisely, under unfavorable conditions (say, stress, sickness, or lack of food), it reverts to the juvenile stage and becomes a polyp again. It can do so even after reaching sexual maturity. *Turritopsis nutricula* isn't indestructible; diseases, predators, and accidents still spell doom. But this creature enjoys biological immorality: for indefinite lengths of time, a process known as transdifferentiation blocks the death of cells and restores those that have suffered damage. Needless to say, research into the system at work holds major prospects for regenerative medicine and cultivating replacement cells in a laboratory setting.

Silently dwelling in the depths of the Caribbean, *Turritopsis nutricula*, like other jellyfish, is thriving. Each year, its numbers multiply, and probably for a number of reasons: its "immortality," global warming, the concentration of CO_2 in the atmosphere, and the overfishing of its principal predators.

Jellyfish, found from the surface of the sea to its depths, are fascinating and terrifying predators. Their shapes, sizes, and colors are sublime, and the larger part of the mysteries they harbor still hasn't been explored.

AN INDESTRUCTIBLE ALIEN

WHAT CREATURES could possibly survive disaster on a cosmic scale, years without water, immense levels of atmospheric pressure, radiation, lack of oxygen, and the quasi-absolute vacuum of space? Do such marvels exist? Yes, they do. They're elementary beings: tiny "water bears," or tardigrades (literally, "slow walkers"). These animals can withstand any trial. They have eight clawed feet and a trunk, and nothing scares them. Resembling arthropods, tardigrades are found in sand, ice, at hot springs near volcanoes, and among the mosses and lichens they feed on (even though they're sometimes cannibals—and eat roundworms by piercing them with their proboscis).

Almost a thousand species of tardigrades have been recorded. They demonstrate a phenomenal capacity to survive intolerable conditions. No individual species possesses every ability, but each one demonstrates exceptional properties. These incredibly small organisms defy death by entering a state of *anhydrobiosis* (or *cryptobiosis*): lowering their metabolism so drastically that they become totally inactive. All the water flushes from their system and is replaced by "antifreeze" (a sugar,

> "And if only one remains,
> I'll be that one!"
>
> —*Victor Hugo*

in fact); simultaneously, a kind of microscopic wax covers their bodies. Then, when circumstances prove favorable again, they restart their metabolism. In other words, tardigrades can revive from "clinical death." Pretty convenient!

There isn't a record tardigrades haven't made or broken. They can endure conditions at −200°C (−328°F); in a laboratory setting, some even made it for twenty hours at about −270°C (−454°F). By all indications, there's a specimen that revived after having spent thirty years on a piece of moss, frozen at −20°C (−4°F)! The capacity to handle the cold doesn't prevent them from surviving at 151°C—or even 360°C (303 to 680°F), for up to an hour. And there's more. After being blasted into space (more than 270 kilometers

above sea level) on the Soyuz rocket and exposed to cosmic radiation, specimens made it back to Earth alive. Plus, certain species can withstand a pressure of 600 megapascals—in other words, conditions at a depth of 60,000 meters! Phenomenal levels of X-rays won't kill them, even a thousand times the lethal dose for a human being. And that's not all! Tardigrades can handle ultraviolet light up to 7,000 kilojoules per square meter (enough to destroy any other organism) and the ionizing radiation produced by nuclear reactors. Need another feat? When difficult circumstances hinder normal reproduction, they turn their gametes (sex cells) into a little egg.

How are these wonders possible? We don't have all the answers. It would be remarkable if science ever fully understood these "water bears." But we do have some leads. Two are especially important. First, it seems that these athletes of survival repair their DNA by means of a protein (*Dsup*, or "damage suppressor"); doing so protects them from X-rays. (Significantly, this protein is transposable to other organisms, including humans.) Second, it seems that tardigrades' powers are linked to genes borrowed from bacteria—which are famously resistant to hostile conditions. And if you like science fiction: some researchers believe such adaptation to extremes could only have been acquired somewhere other than our planet—and that tardigrades came to Earth on a meteorite about 90 million years ago!

So are tardigrades indestructible aliens? Whatever they are, they're clearly good at hanging in there …

HIBERNATUS

DURING THE LONG, harsh winter, the wood frog lies immobile in the earth. It's literally frozen. But when spring returns, it comes back to life, having survived temperatures as severe as −20°C (−4°F), which kills other species. We human beings have difficulty freezing organic matter at this level. So how can a frog do it?

Let's take a closer look. The wood frog burrows into the forest floor in order to make it through the winter—which, needless to say, can be pretty rough in Alaska. Once it has done so, it suspends its vital functions: its breathing ceases, and its blood barely circulates. The frog's heart has practically stopped, and its legs are brittle. All the same, it's alive! This is possible because glucose from its liver keeps its blood and cells from freezing. Production of this sugar is triggered when the frog's skin starts getting really cold. The glucose has cryoprotective properties; in other words, it prevents the cell degradation that occurs when water freezes—and, conversely, when the crystals start to thaw.

Usually, the process of freezing destroys cells from the inside, causing the body to die. But in the wood frog's case, freezing takes place between cells; as such, it doesn't rupture the cell membranes.

Another element helps it survive the intense cold: urea, which also has cryoprotective qualities. The Alaskan wood frog has a concentration of this substance that's three times greater than the quantity found in other frogs.

The Alaskan wood frog hasn't revealed all its secrets yet. Its abilities, unbelievable to human beings, are the stuff of dreams—and research!

CAMOUFLAGE,
PROTECTION,
DEFENSE

INSECT OR LEAF?

THE GIANT LEAF INSECT
Phyllium giganteum
Size: 5–12 cm

THIS ANIMAL is a stick insect, a master of camouflage that practices its art so perfectly that, in imitating a leaf, it shows signs of "damage": little brown spots, hints of veins, and even the irregular and withered edges found on aging foliage. Nor is that all: it sways back and forth to mimic a leaf blowing in the wind. A leaf with legs! The degree of realism is stunning. Such mimicry is vital for fooling predators—especially for females, whose *elytra* practically cover their body and prevent flight. During the day, these creatures are largely immobile; under the cover of night, they move slowly and feed on guava or mango leaves.

Besides their impressive camouflage, giant leaf insects display obvious sexual dimorphism. Females are stocky, whereas males are long and thin. Females have finer antennas, too. (The antennas of males likely serve to capture pheromones emitted by females.) They reproduce sexually, but a good man can be hard to find. We don't even know what the males of certain species of leaf insects look like! Plus, they're known for occasional thelytokous parthenogenesis: unfertilized females can lay eggs that hatch only female young; the new generation will also become one with the surroundings, thanks to the art of illusion. Doing so means that they'll be less likely to be eaten—so they can pass their "mimetic genes" on to their own young.

"On this island, we also found a tree whose leaves, when they fall, become animated and walk about."
—Pigafetta

CATERPILLAR OR SNAKE?

THE SPHINX MOTH
Hemeroplanes ornatus
Size: 4–7 cm

THE LARVA of the sphinx moth enjoys a phenomenal ability. When a predator approaches, it puffs up its head and chest. Just like that, it transforms into a tree viper, imitating the snake's scales and triangular head. The reptile it mimics is poisonous and to be avoided. To enhance the effect, the caterpillar expands part of its thorax and displays marks that look like two eyes. In this way, it convinces attackers that they're facing a snake—much tougher prey, in other words.

The impersonation isn't limited to physical resemblance. The caterpillar of the sphinx moth goes further still. With uncanny accuracy, it copies the way the snake moves when it's about to strike, rearing up and swaying with the front part of its body. The eyespots on display in the defensive position ensure maximal protection from predatory birds, even at latitudes where tree vipers aren't particularly common. Remarkably, some species closely related to *hemeroplanes* (which don't look as convincing) have real venom, or even a small appendage that darts out from a little opening, just like a snake's tongue! But whatever tricks are used, these caterpillars' complex subterfuge acts as a deterrent: birds get scared and head off to find an easier meal.

> "To normal people this caterpillar
> might look weird and scary,
> but for me it's just a walk in the
> park. Every caterpillar in Costa Rica
> looks like something else."
>
> —Daniel Janzen

Ironically, the down of chicks belonging to one kind of predator (*Laniocera hypopyrra*) looks just like the stinging bristles of a big, fat caterpillar! To enhance the illusion and trick potential attackers, the down undulates, just like the spines on the animal being imitated. Mimicry can take anyone as a model—including the masters of mimicry themselves.

THE LITTLEST SEAHORSE

THE PYGMY SEAHORSE
Hippocampus bargibanti
Size: 0.5–2.5 cm

THE PYGMY SEAHORSE is one of the smallest members of its family. It lives among soft corals and seagrasses. Simply moving vertically, by means of its single dorsal fin, proves exhausting; in consequence, it's relatively slow. This tiny creature has no protective scales, claws, or venom. Its chief resource is an incredible gift for mimicry. Indeed, pygmy seahorses live in a kind of symbiosis with gorgonians (*Muricella sp.*), to which they cling with their prehensile tails—as if they were extensions of them! Pink, red, or orange nodules cover their body and make them blend in; it's hard to find a better example of using mimicry to avoid predators.

Being undetectable is fine, but one still needs to eat. Indeed, the pygmy seahorse has to eat frequently, because it barely has a stomach. To compensate for its slow rate of movement, this little carnivore possesses a tubular mouth into which it sucks prey unlucky enough to come near: crustaceans, larvae, eggs, and other kinds of zooplankton. The pygmy seahorse hunts by stealth. It stabilizes itself by means of its tail and, thanks to eyes that move independently, enjoys a panoramic vision of the surroundings. Of course, once it has found safety and been fed, it also needs to reproduce. Pursuant to a nuptial dance performed by the male, partners twist their tails around each other and mate. Remarkably, it's the male that carries the eggs—in a stomach pouch where the female has deposited them—until they hatch and baby seahorses emerge.

"One day, an earthquake raised the sea and submerged the whole city, including the temple of Poseidon. … Eratosthenes says that he saw the place, and the ferrymen say that Poseidon's bronze statue stood upright in the strait, with a seahorse in its hand."

—Strabo

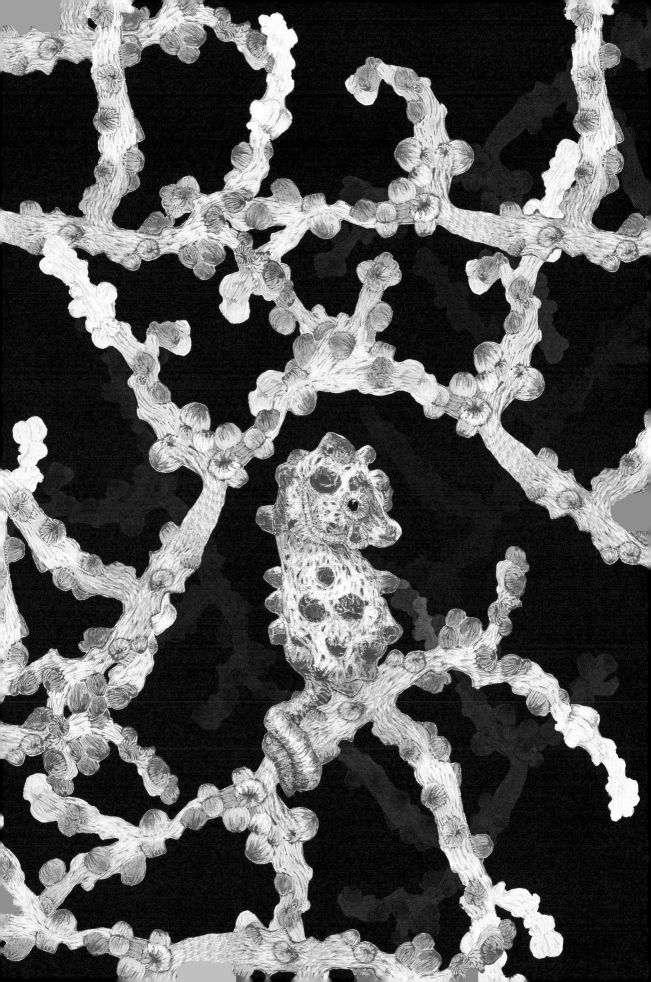

FATAL BEAUTY

THE BLUE POISON DART FROG
Dendrobates azureus
Size: 3–4.5 cm

THE BLUE POISON DART FROG lives in savannahs and forests, especially on moss and rocks near water. For the most part, it prefers the ground, but it can climb trees easily. Its diet consists of assorted insects and arthropods: ants, termites, crickets, fruit flies, spiders, springtails, and so on. Petite and bright blue, with black spots unique to each individual, this frog manages to be cute and exquisite at the same time. But it's unwise to trust appearances … or, then again, maybe it isn't. When frogs are extremely colorful, it's a bad sign—at any rate, it's a warning. This frog is poisonous (not venomous—it doesn't actively introduce toxins into other creatures), and its bright color tells anybody planning to eat it to back off. The scientific term for this strategy is *aposematism*.

Once they've made the mistake of trying a taste, predators remember that the frog is dangerous and stop attacking it—the risk is just too high. These amphibians secrete mucus with alkaloid toxins that act on the nervous system and paralyze the aggressor's body. The frog's own poison is formed from toxins in the termites or ants it has consumed; the process compounds their potency. Indeed, it seems that the frog's mucus alone can kill a human being. Some cousins of the blue poison dart frog are even deadlier. A single gram of poison from the world's most toxic frog species (*Aparasphenodon brunoi*) can kill three hundred thousand mice—or eighty people! Evidently, it's the yellow-banded poison dart frog that inspired Amazonian Indians to use poison to coat their arrows.

> *"Grief—that's beauty's canker."*
> —*William Shakespeare*

THE BLUE ANGEL

THE BLUE DRAGON SEA SLUG
Glaucus atlanticus
Size: 3–6 cm

GLAUCUS ATLANTICUS is a strange beast. A gastropod mollusk, it belongs to the family of nudibranchs, which means "naked gills." Another name for these beautiful animals without shells is "sea slug." This particular species is also called "blue angel" or "blue dragon," because it resembles a creature of myth. It's equipped with conical growths (*cerata*) grouped in clusters, like bouquets, which serve as gills.

This magnificent little creature moves slowly and in rudimentary fashion. It has no shell, and its sensory capacities aren't very developed. It doesn't spit fire, either. For all that, the blue dragon sea slug is quite formidable. Indeed, it sometimes feeds on creatures larger than itself: cnidarians floating on the surface such as *Physalia physalis* (Portuguese man-o'-war), *Porpita porpita*, and *Velella velella*. Its nematocysts contain stinging—and highly toxic—substances. The way the blue dragon makes use of this feature is absolutely fascinating, and it represents an extraordinary evolutionary adaptation. In fact, the blue dragon gets the nematocysts from its prey, storing some of them, undigested, on the outside of its body.

"This charming animal … has surely struck all naturalists to have sailed the sea, by the elegance of its form and its bold and agreeable assortment of colors."

—*Georges Cuvier*

The animal itself isn't affected because specialized cells in its epithelium (skin) form a layer that also lines the mouth; its secretions protect it from its prey. When it encounters a new kind of poison, the chemical composition of the mucus changes! *Glaucus atlanticus* itself is immune—but other animals aren't; in the event of contact with a predator (or human being), the nematocysts release their powerful venom, resulting in anaphylactic burns and shocks.

This defense strategy is bolstered by the blue dragon's movement: thanks to the air in its stomach, it drifts upside-down, on its back, and on the water's surface, following the wind and currents. What, you ask, is the point of this seemingly ridiculous way of swimming? The blue dragon benefits from a kind of camouflage also used by sharks, counter-illumination. The animal's ventral side is blue and white, which hides it from aerial predators; in turn, the dorsal side is silvery gray, which makes it hard for marine predators to detect. All the same, sea dragons can fall victim to other carnivorous nudibranchs, fish, crabs, spider crabs, or marine worms. But there's no risk of them disappearing. Their mode of reproduction just might make all the difference: the blue dragon is hermaphroditic. Each individual has male and female reproductive organs. Incapable of self-fertilization, partners mate face to face, holding themselves in place with their penile hooks and *cerata*. Each one then lays hundreds of eggs at sea, near the ground, in mucus in the form of ribbons or lace, or on the carcass of a victim.

Its bright colors and lateral excrescences give the blue dragon sea slug an appearance that is symmetrically pleasing and unusual.

AN ARMORED DEVIL

THE THORNY DEVIL
Moloch horridus
Size: 15–20 cm

THE THORNY DEVIL, or simply "moloch," is a solitary lizard that lives in the central Australian desert. No one can fail to be astounded by this marvel. Its exterior morphology is simply incredible. With his rhinoceros-like head and spiky armor, it seems to have come straight out of a science-fiction movie about dinosaurs.

And yet this creature is real. Equipped with sharp spines all over its body—in particular, scaly plates on its head and back—it has all it needs to frighten off its main predators, birds and snakes. Not to mention the thorny ball behind its head, which makes it look like it has a second head—a good trick for fooling attackers. And as if that weren't enough, the thorny devil's behavior complements the strength of its armor: when threatened, this creature puffs up to look even more intimidating. Only monitor lizards and some birds of prey aren't impressed. The armor affords protection against most attacks, plus it provides camouflage in the animal's natural environment—the desert—thanks to its gray, beige, and orange hues. To strengthen the effect, the thorny devil moves in spurts, interrupted by abrupt stops; this makes it even harder to see.

The moloch has other tricks, too. It feeds exclusively on ants, by means of its darting and deadly accurate tongue. It's an excellent hunter with digestive capacities second to none: a single individual can devour three thousand ants in just one meal! Another impressive feature is how the spines serve as "straws" for drawing moisture from the humid soil or water that falls from plants onto the animal's back; by capillary action along a complex network, the vital element winds up in the moloch's mouth. It's quite the clever adaptation to life in the desert, where water is so scarce.

Between its perfect camouflage and uninviting appearance, the thorny devil has done everything possible to survive.

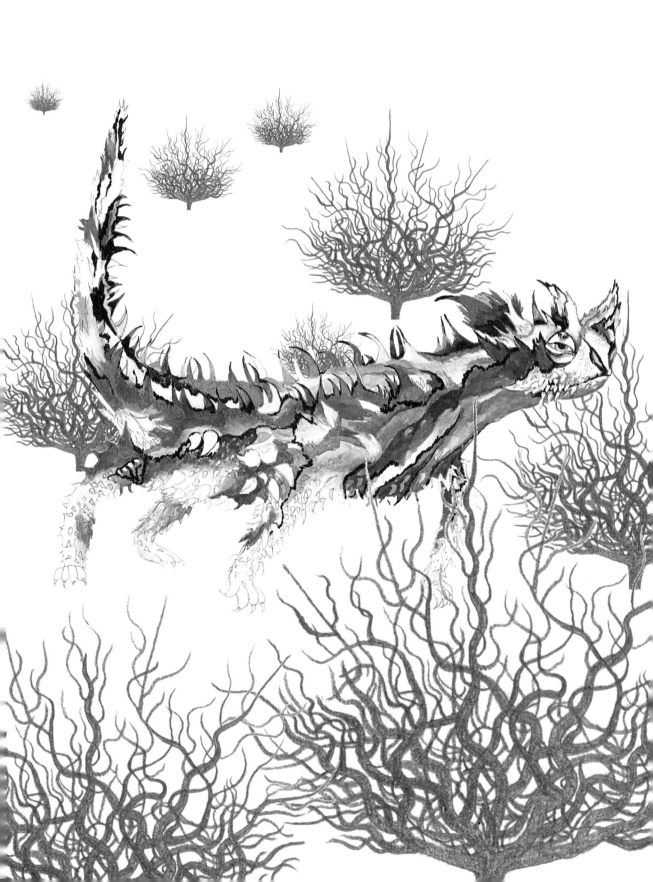

A FEARLESS TERROR

THE HONEY BADGER
Mellivora capensis
Size: 50–80 cm

THE HONEY BADGER, also known as the "ratel," is a small, solitary carnivore. It loves honey, but it also feeds on reptiles, termites, scorpions, and earthworms that it digs up with claws that are four centimeters long; it also pursues large prey such as porcupines and hares—and even wildebeest and antelopes. This creature's a real terror. No animal can scare it, not even dangerous ones like lions, jackals, hyenas, cheetahs, and buffaloes. Its toughness and aggression have made the honey badger a legend of the savannah.

The honey badger considers deadly snakes like puff adders (*Bitis arietans*) and Cape cobras (*Naja nivea*) delicacies! How is this even possible? After taking a bite, this remarkable animal falls into a coma for a few hours—long enough to metabolize the venom. Then it finishes eating the snake and goes off to find another one! Honey badgers are taught to hunt by their mothers, and they acquire immunity from poison by exposure to snakes and scorpions at a very young age. They also have an array of defensive strategies—the best of which, of course, is attack. A honey badger faces its enemy down, no matter what, offering resistance while disorienting the other animal by performing a kind of dance—one step back, then two steps forward. Plus, it secretes a nauseating odor. In the rare event of capture, the honey badger won't give up without a fight; immediately, it strikes at the enemy's scrotum to cause a hemorrhage. It also has incredibly tough skin; even lions have trouble piercing it.

Small but stocky, ingenious, and courageous (or reckless), up to any challenge, the honey badger is an inspiration even for rugby players!

In captivity, the honey badger's a real Houdini. In South Africa, these animals have been known to pile up stones, rakes, tires, and other objects to climb over fences; they also dig tunnels and even pick locks—a nightmare for zookeepers!

SURVIVING THE COLD, UNDERGROUND, AND IN THE DESERT, SKY, AND WATER

THE EMPEROR OF THE ICE

THE EMPEROR PENGUIN
Aptenodytes forsteri
Size: 1.15–1.25 m

SURVIVAL IN THE FREEZING Antarctic is a big challenge, even for this streamlined bird whose body is perfectly adapted to swimming and diving. In fact, the largest emperor penguins can stay underwater for over twenty minutes, at a depth of more than 500 meters, thanks to hemoglobin that functions even with minimal oxygen and a skeleton that's built to resist pressure. This allows them to hunt fish, crustaceans, and cephalopods. But with an air temperature that can reach just about −50°C (−58°F), an icy wind blowing at nearly 150 kilometers per hour, and seawater at a temperature of almost −2°C (28.4°F), these animals need to survive the cold, above all.

"To start again—where the world began. I'll go to dream in that candid paradise, where emperor penguins frolic at sunrise, showing us what it means to be alive. I'll go dream in that candid paradise, where the air is so pure one can bathe in it, and play in the wind, as in the dreams of my childhood, as before."
—*Michel Berger*

This majestic bird has a veritable arsenal for fighting the temperature and limiting its own loss of heat. First, its feathers: penguins have the densest plumage of any bird; it covers almost the whole body and provides more than 85 percent of its insulation. Specialized muscles permit them to orient their feathers so they can maintain a layer of air on the skin, which promotes the conservation of warmth. To add further insulation and protection from the hostile surroundings, dense down forms a blanket at the base of the feathers. Plus, emperor penguins have a layer of fat three centimeters thick. In the event of major thermal distress, a process of homeostasis occurs under the impetus of two hormones: insulin and glucagon. Even when it's −47°C (−52.6°F) out, emperor penguins can maintain a body temperature between

37.5°C and 38.5°C (between 99.5 and 101.3°F). Besides these resources—plumage, fat, and hormones—penguins have a perfectly adapted vascular system: their blood viscosity increases when the temperature's low, and peripheral vasoconstriction makes it possible to limit thermal losses in tissue exposed to cold. Finally, a system of "recycling" air at the level of the nasal cavities prevents excessive loss through evaporation.

Appropriate behaviors complement these anatomical adaptations. Emperor penguins regularly groom their plumage, coating it with secretions made of fatty substances and waxes from their uropygial gland in order to provide the impermeability essential for good insulation. Plus, they move a great deal to keep warm, swimming, walking (with or without an egg or chick), or simply shivering. Finally, they practice the sensible and effective strategy of social thermoregulation. In response to extreme temperatures and wind conditions, individuals band together to benefit from the heat of the group. Doing so limits thermolysis, the loss of thermal energy to surroundings. The goal of penguins—the only animals to breed during the Antarctic winter—is to conserve the energy necessary for incubating their eggs. How do they manage that? First, they group into a cluster called a "huddle"! Birds playing football? The penguin huddle is a dense, defensive formation—not against enemies, but against the elements. Thousands of individuals—between eight and ten per square meter—can squeeze together, yielding a temperature up to 37.5°C (99.5°F)!

One might assume that the biggest and brawniest would push to the middle, relegating the younger and weaker to the sides. In fact, however, emperor penguins aren't very aggressive, and it seems that access to good positions is equitable: according to need, individuals have the opportunity to bunch up or not. As a result, during the incubation period, the formation proves quite varied; all birds can benefit from the warmth of the group. In turn, when fierce blizzards or glacial winds prevail—or when it's time to mate—

penguins at the edges move toward where it's sheltered, thereby causing the most coveted positions to change occupants. The exact dynamics of social thermoregulation are very complex and still have not been fully elucidated.

A universe of ice, whiteness, and cold—and yet, so warm, too. A regal bird is passing us the torch of its beauty and the fierce struggle for survival, in a fantastic setting that inspires the dream that we, too, may yet live in a world so immaculate.

A TENTACULAR BURROWER

THE STAR-NOSED MOLE
Condylura cristata
Size: 16–24 cm

MOLES HAVE TO SURVIVE underground. Among their number, the star-nosed mole (*Condylura cristata*) is especially fantastic, like a creature somebody dreamed up! A semiaquatic animal that's active both day and night, it lives in colonies. Star-nosed moles construct their burrows from vegetable matter; often, their homes include passages opening directly onto streams or ponds. During the winter, they have no trouble digging under the snow or diving below the ice. The star-nosed mole is an accomplished swimmer and enjoys feeding on invertebrates: mollusks, worms, and even fish, which it detects by sound and smell, as its eyes are tiny. Emitting an unappetizing odor, it fears few other animals, even though large predators such as raptors and pike will sometimes attack. In addition to its scent, the star-nosed mole keeps itself safe by digging underground tunnels up to 270 meters long.

This animal has everything it needs to excel in its environment. Its front limbs are particularly well-adapted to digging: they're short and muscular, and they end in two big paws equipped with mighty claws facing outward. In addition, the mole has a thin coat that can smooth out in any direction; this feature helps it move backward in its narrow burrow. Although its eyes are small—the star-nosed mole is practically blind—it has a keen sense of smell, even in the water. Its sense of touch is also excellent, among the best of all mammals: a long muzzle with twenty-two tentacles enabling it to probe sediment underwater and the walls of its subterranean tunnels. When the mole digs, it folds these feelers over its nostrils to protect them from the dust. The appendages are always moving at an exceptional speed; they're lined with thousands

The bizarre star-nosed mole, with its octopus-shaped snout, looks scary. But it's also fascinating, and we still don't know all its secrets.

of nerve fibers and tens of thousands of mechanical receptors (known as Eimer's organs). Indeed, the star-nosed mole can even sense the electrical field generated by a real delicacy: earthworms. In less than a quarter of a second, it locates and seizes its prey; the feat qualifies it as one of the world's fastest eaters. (The worms panic and try to rush to the surface whenever a mole shows up.) And there's more. It seems the tentacles have two functions: longer ones serve to locate prey at a distance, and shorter ones are for inspection at close range.

"How can they feel detailed textures with a single touch of their rays? What genes and molecules allow the star to develop, and how does its brain so greatly amplify the touch signals coming from its nose? The mole doesn't hibernate in the winter, so how does it keep its sensitive star working when it dives into ice-cold water?"
—Ken Catania

THE KILLER
OF THE BURNING SANDS

....

THE HORNED DESERT VIPER
Cerastes cerastes
Size: 60–80 cm

....

SURVIVING IN THE DESERT … that's life for some snakes. Though most of their kind are harmless, this isn't true of the horned desert viper, one of the most dangerous desert animals to be found. Bearing two pointy scales on its head, it delivers venom by means of an elaborate series of mobile hooks that reach deep into its victims' flesh. Don't worry: this creature doesn't go around killing systematically. Sometimes its venom, which acts as an anticoagulant, only causes hemorrhagic edema. The horned viper is ovoviviparous, which means that it doesn't lay eggs but gives birth to live young. Beautiful and elegant, it represents a crowning achievement of adaptation to life in the desert environment.

The horned viper slides gracefully over the sand; sometimes, it hides there too, both to protect itself and, especially, to lie in wait—like any good opportunistic predator. The color of its skin blends in with the sand of the deserts of the Sahel and Sahara, as well as the Arabian Peninsula, where it hunts rodents, lizards, arthropods, and insects. The patterns on its body and the color of its iris imitate the nuances of the landscape perfectly. All but invisible, it moves slowly and sinuously so no one will detect it. To avoid sinking, it rests on two points of its body at once; the rest launches forward in a series of loops. In this manner, the horned viper can advance on the soft surface and, at the same time, regulate its body temperature by minimizing contact with the hot ground. The contractions and contortions of its body, which is muscular

"When I was just a kid … my curiosity about snakes got the better of my fear. … Now I'm working to change the reputation of these unloved creatures. I try to show that they're absolutely marvelous animals … whose biology is astonishing. I've spent thousands of hours on the field trying to turn myself into a tree trunk … just to observe them."
—Françoise Serre Collet

and flexible, allow it to move sideways over the dunes. When the sand is particularly hot, this snake can advance very quickly—by leaping—which also enables it to climb steep inclines.

Another adaptation to parched and stifling conditions: when heat is most severe (around 44°C or 111.2°F), the horned viper burrows so that only its eyes are exposed. In so doing, the snake can lower its internal temperature by 10°C (18°F) and "simply" survive. By the same token, it adjusts its behavior to the season, becoming active only by night in summer and during the day in winter. To be sure, its skin is especially adapted to the task of conserving as much water as possible. A lipid layer surrounded by keratin provides perfect insulation. What's more, the quantity and arrangement of these lipids correlate with the dryness of the surroundings; the snakes are even capable of adjusting them between molts. Finally, their kidneys are made to filter the maximum quantity of water. As a result, their urine is solid, composed primarily of uric acid. The horned desert viper, which has greater need of water than forest snakes, literally urinates pebbles! Some people believe that when there's a drastic shortage, this snake can extract water from its venom. This hypothesis has yet to be verified, however, and the precise mechanism isn't known. But if this is the case, the venom would become even more concentrated and dangerous—strong enough to kill a dromedary or human being in ten minutes! Is it just a myth?

Despite their beauty, elegance, strength, and unbelievable physiology, snakes inspire fear—if not disgust. Each year, there are two million cases of venom poisoning, with 20,000 to 24,000 fatalities. But it's important not to forget the main thing: only 10 percent of snakes are poisonous to human beings, and their population is declining all over the world. What's more, snakes are fascinating, and much more remains to be discovered about them— especially their intelligence.

A BIRD OF THE HIGHEST ORDER

···

WILSON'S BIRD-OF-PARADISE
Cicinnurus respublica
Size: 16–21 cm

···

SURVIVAL IN THE AIR: birds have a wide range of resources for making the best of it. Life in the air means taking off, flying, soaring, migrating, landing again … in certain cases, on the water. Birds have mastered all these skills. To pay due honor to all their accomplishments, what's more fitting than to evoke the glorious species known in French as the "republican bird-of-paradise"? This forest-dweller feeds on arthropods and fruit. Males are famous for their courtship display, when they dance and exhibit their dazzling colors. But our interest here doesn't bear on courtship so much as flight.

First and foremost, Wilson's bird-of-paradise is a bird, and being a bird (generally) means flying. Flying means taking off. This, one assumes, is a matter that primarily involves wings. In fact, however, the most important thing for launching into the air isn't wings, but legs. The legs enable birds to build up speed. Other types of take-off exist: birds can make an extended run on the ground or water, launch straight up, and also start with or without the help of the wind; some take a plunge into the void from a branch. What-

> *"If we don't have the paradise of a republic, there should at least be a republican bird-of-paradise!"*
> —*Charles-Lucien Bonaparte*

ever the method, once the bird's no longer in contact with the ground and has overcome gravity, the wings take over. Now, the point is to keep flying and not to fall. On this score, the republican bird-of-paradise displays a vigorous, undulating pattern of flight. A consummate creature of the air, it has a perfectly adapted body: a light skeleton (thanks to hollow bones), aerodynamic contours, plumage that ensures lift, warm-blooded circulation that allows a constant flow of energy, a brain and organs that have evolved to sense air currents, and the ability to use those currents to change direction by moving its pinions. And that's not including the complex adaptations it shares with all migratory birds, which

compensate for weight loss on long journeys and help navigation under adverse conditions. (In this regard, the resources available to the homing pigeon—which is not in the same league as our bird-of-paradise in aesthetic terms—still haven't been fully explained.)

Once aerial feats are completed, birds must land. Depending on the species, this occurs on a branch, a flat surface, a rock, or the water. Needless to say, each kind of bird has its own technique. The bird-of-paradise basically lands on branches, and the shape of its claws guarantees a perfect grip and stability. But wherever birds come back down to earth, they demonstrate a few points in common. Feathers always spread as the approach is made, in order to increase lift. Additionally, birds slow down by increasing the angle of incidence—like a plane! Finally, the tail feathers soften the moment of contact.

Wilson's bird-of-paradise—especially the male—is extraordinarily beautiful. Its striking looks and majestic display are almost enough to make one forget that every day, like most other birds, it performs the wonder of flight, too.

The natural scientist Charles-Lucien Bonaparte came up with the name for this bird in French. As the nephew of Napoleon Bonaparte, he loathed the widespread habit of his contemporaries to give royal names to new species. In this way, he honored the republican cause and thumbed his nose at the monarchy!

A WONDER
OF THE MARINE DEPTHS

THE GOAL OF AQUATIC ANIMALS is also survival. One of them has no equal when it comes to adapting to the marine environment. The southern blue-ringed octopus is a benthic octopod (in other words, an eight-legged dweller of the seafloor) inhabiting coral reefs, where it feeds on crustaceans, mollusks, and small fish. Its evolution allows it to breathe, move, hunt, and defend itself perfectly in these surroundings.

To breathe in the water, octopuses have gills—like fish—but they're shaped like feathers and hidden in a muscular cavity whose opening lies behind the tentacles. These creatures suck in water, and their gills capture dissolved oxygen. To diffuse the oxygen, octopuses have three hearts: two that move blood toward the gills, where it loads up on oxygen, and one so-called systemic heart that pumps the oxygenated blood to the rest of the body. Finally, a mobile siphon located under the belly expels the water. When this happens quickly, it produces vigorous propulsion, which helps the animal escape predators. Subtler movement takes place through undulating fins, which are assisted by the tentacles; in this way, the octopus crawls on the seabed to catch food, which it shreds with an extremely hard and sharp beak.

Cephalopods have no reason to be jealous of vertebrates; they're far from having revealed all their secrets and still inspire dreams and flights of imagination.

That said, the blue-ringed octopus excels most when it comes to self-defense. First of all, millions of pigment cells ("chromatophores") give it—like many other cephalopods—the power of *homochromy*, the ability to change color in order to blend in with surroundings. Some octopuses even use shells for cover, or put their tentacles over the gills of predators like sharks in order to stifle the attack and escape. Besides sophisticated camouflage, which sometimes means assuming the appearance of other species (certain cephalopods

disguise themselves as hermit crabs), octopuses use clouds of black ink, which they project while speeding off, in order to gain time to hide. And if that's not enough, the blue-ringed octopus will bite, injecting a venom in its saliva that's one of the most powerful in the animal world. The venom contains tetrodotoxin, the neurotoxin we saw in the pufferfish, which causes death by respiratory failure in a matter of minutes; it's strong enough to kill a human being. In other words, don't mess with the blue-ringed octopus …

To survive in an aquatic environment, cephalopods have three hearts, millions of neurons, an unusual sensory system—they gauge water pressure by smell—tactile organs that boast thousands of autonomous suction cups, vision that registers how surfaces refract light, and the ability to live in colonies. Since mothers die shortly after eggs hatch, it might seem that these creatures can't transmit the knowledge and skills they've gained. But now we know that cephalopods memorize, learn, invent, and play; indeed, they acquire information about the outside world while still in their eggs!

*"Not only is it supposed that these poulps can draw
down vessels, but a certain Olaus Magnus speaks of
an octopus a mile long that is more like an island than
an animal. It is also said that the Bishop of Nidros
was building an altar on an immense rock. Mass
finished, the rock began to walk, and returned to the
sea. The rock was a poulp."*
—Jules Verne

CONCLUSION

The animal world today reflects a lengthy process of evolution, which is ongoing; its mechanisms are complex, the adaptations infinite. All in all, we know only a minuscule part of this world—both in terms of species (some 90 percent probably remain unknown) and in terms of animals' amazing abilities, many of which still await discovery. Understanding how the adaptations they've developed interact with their environments, on a dizzying scale that reaches back millions of years, is a continuous source of fascination, almost a waking dream that never leaves me.

This book has surveyed a microscopic portion of the animal world—just a fraction. The material already defies norms of classification, reproduction, behavior, physiology, and anatomy. But then again, doesn't every species—each individual creature—now defy our world, which is changing at a breakneck pace? What treasures still lie in store? What remarkable abilities? What beauties? Animals harbor infinite surprises, if only we give them time to reveal them. There's a whole universe awaiting discovery, but we need to act fast. Very fast. Eighty percent of the insects in Europe have vanished—along with 30 percent of the birds that rely on them to exist. I'm not done marveling at the masterpieces nature offers. I want my little boy to experience the same wonder I did when, as a child, I watched ladybugs in my grandmother's garden. Whether it's a tiny spider or an elephant, every life has value. But we no longer feel its value. The situation is infinitely sad. Disgusting. Overwhelming. Intolerable. I'm distressed by the tragedy that has occurred during my own short life. My words are like tears. "It's a sad thing to think that nature speaks and mankind doesn't listen," Victor Hugo wrote.

I feel that there's nothing more important than to pass on, to my son, the little piece of nothing and everything that I've observed—the happiness that comes from watching a dragonfly, spider, frog, lizard, elephant, parrot, mouse, orangutan, or ladybug ... it's the joy of witnessing life itself. No excursion into nature, not a single one, has failed to astonish and move me. The experience brings laughter, tears, joy, fear, puzzlement, admiration, wonder, and tenderness. Each individual creature enriches my own existence boundlessly. Extraordinary beings populate the land, air, and sea. One just needs to take the time to look. Open your eyes! Behold the world and all the life it contains. Look at *them*. Let yourself be moved. Human beings are capable of great things—very great, sometimes. I can still hear the heartbeat of life just three weeks old and a little bigger than a millimeter in my womb … wondrous technology made it possible for me to hear it. Humankind in always envisioning, inventing, and sharing. We should draw inspiration from this enduring vitality, which has weathered so many crises of extinction. Life has always resisted death. We must act. Respectfully, humbly, and creatively, we must act. I'm smiling as I write these words. Once upon a time, a dream … or two … Once upon a dream … Let's make the dream last.